BETTING ON FAMINE

BETTING ON FAMINE

Why the World Still Goes Hungry

Jean Ziegler

Translated from the French
by Christopher Caines

THE NEW PRESS

NEW YORK
LONDON

The New Press gratefully acknowledges the Florence Gould Foundation for supporting the publication of this book.

Originally published in France as *Destruction massive: Géopolitique de la faim* by Éditions du Seuil, Paris, 2011
Published in the United States by The New Press, New York, 2013
Distributed by Perseus Distribution

LIBRARY OF CONGRESS CATALOGING-IN-PUBLICATION DATA

Ziegler, Jean.
  [Destruction massive. English]
  Betting on famine : why the world still goes hungry / Jean Ziegler ; translated from the French by Christopher Caines.
      pages cm
  Includes bibliographical references.
  ISBN 978-1-59558-849-4 (hc. : alk. paper) -- ISBN 978-1-59558-861-6 (e-book) 1. Food relief--Political aspects. 2. Hunger--Political aspects. 3. Food and Agriculture Organization of the United Nations. I. Title.
  HV696.F6Z5413 2013
  363.8--dc23
                          2013005169

The New Press publishes books that promote and enrich public discussion and understanding of the issues vital to our democracy and to a more equitable world. These books are made possible by the enthusiasm of our readers; the support of a committed group of donors, large and small; the collaboration of our many partners in the independent media and the not-for-profit sector; booksellers, who often hand-sell New Press books; librarians; and above all by our authors.

www.thenewpress.com

Composition by Bookbright Media
This book was set in Baskerville and Franklin Gothic

Printed in the United States of America

10 9 8 7 6 5 4 3 2 1

CONTENTS

The man who wants to keep faith with justice
must ceaselessly break faith with injustice
in all its inexhaustibly triumphant forms.
—Charles Péguy

ACKNOWLEDGMENTS

Erica Deuber Ziegler collaborated closely with me on the development of this book. With infinite patience, great savoir faire, and unfailing erudition, she read, edited, and reorganized all ten successive versions of the manuscript. Olivier Bétourné, the president of Éditions du Seuil, conceived of the book, personally edited the final version, and came up with the title. His stimulating friendship has been of decisive help to me.

My colleagues on the Human Rights Consultative Committee of the UN Human Rights Council (UNHRC), Christophe Golay, Margot Brogniart, and Ioana Cismas, assisted me in researching the book. Sustained by our shared convictions, both their indefatigable commitment and great professional skill have been indispensable.

James T. Morris, Jean-Jacques Graisse, and Daly Belgasmi opened the doors of the World Food Programme to me. Jacques Diouf, director general of the WFP, and many of his colleagues were generous with their assistance.

Pierre Pauli, a statistician in Geneva's Cantonal Statistical Office, helped me to master the crushing mass of data on hunger and malnutrition.

At the UN's Office of the High Commissioner for Human Rights (OHCHR), I have been fortunate to be able to rely upon the skillful, discreet, and sound advice of Eric Tistounet, chief of the OHCHR HRC Branch.

Beat Bürgenmeier, dean emeritus of the Faculty of Economic Sciences at the University of Geneva, and banker Bruno Anderegg initiated me into the complicated world of stock exchange speculation and hedge funds.

Francis Gian Preiswerk was for seventeen years one of the most famous traders at Cargill, a multinational giant in the food services industry. He welcomed me for many in-depth discussions, kindly read selected chapters of this book—and in irate letters expressed his total disagreement with practically all my arguments. However, his rich experience in trade, his exceptional professional abilities, and his generous friendship have been for me beyond price.

With exemplary care, Arlette Sallin prepared clean copy for each successive version of the book; generous with both her time and her well-informed critique, she has accompanied me throughout my work on this project. I was fortunate also to enjoy the advice of Sabine Ibach and Vanessa Kling. Hugues Jallon, editorial director of the social sciences division at Éditions du Seuil, as well as Catherine Camelot, also offered valuable assistance.

To all, I express my profound gratitude.

LIST OF ABBREVIATIONS

AGRA	Alliance for a Green Revolution in Africa
AQIM	Al-Qaeda in the Islamic Maghreb
ASCOFAM	World Association for the Struggle Against Hunger
AU	African Union
CCC	Clean Clothes Campaign
CIAN	Conseil Français des Investisseurs en Afrique (French Council of Investors in Africa)
CODEN	Comité de Développement de la Région de N'do (Committee for the Development of the N'do Region), Cameroon
DPKO	UN Department of Peacekeeping Operations
DRC	Democratic Republic of the Congo
ECOSOC	UN Economic and Social Council
FAO	UN Food and Agriculture Organization
FCI	Food Corporation of India
FIAN	FoodFirst Information and Action Network
FINMA	Finanzmarktaufsicht (Financial Markets Authority), Switzerland
G20	Group of Twenty Finance Ministers and Central Bank Governors
G8	Group of Eight (and similarly: G6, G8+5)
GATT	General Agreement on Tariffs and Trade
GBD	Global Burden of Disease
GDS	Grands Domaines du Sénégal (Large Estates of Senegal)
ICDs	Integrated Child Development Centers, India
ICRC	International Committee of the Red Cross
IFAD	International Fund for Agricultural Development
IFPRI	International Food Policy Research Institute
ILO	UN International Labour Organization
IMF	International Monetary Fund
IRFED	Institut International de Recherche et Formation Éducation et Développement (International Institute for Research and Training in Education and Development)
LAP	Libyan-African Portfolio, Mali
LDCs	least developed countries
LEPI	Liste Électorale Permanente Informatisée (Computerized Permanent Electoral List), Benin
MDGs	UN Millennium Development Goals
MST	Movimento dos Trabalhadores Rurais Sem Terra (Landless Rural Workers' Movement), Brazil

NGO	nongovernmental organization
NKVD	Narodnyy Komissariat Vnutrennikh Del (People's Commissariat for Internal Affairs), the regular and secret police force of the USSR under Stalin; its successor was the KGB
OECD	Organisation for Economic Co-operation and Development
OHCHR	UN Office of the High Commissioner for Human Rights
OIC	Organization of the Islamic Conference
OIP	UN Office of the Iraq Programme
PDS	Public Distribution System (for food), India
PT	Partido dos Trabalhadores (Workers' Party), Brazil
RNS	Reichsnährstand (Reich Food Corporation), Nazi Germany
ROPPA	Réseau des Organisations Paysannes et des Producteurs d'Afrique de l'Ouest (Network of Farmers' and Agricultural Producers' Organizations of West Africa)
Socapalm	Société Camerounaise de Palmeraies (Cameroon Palm Plantations Company)
SOMINA	Société des Mines d'Azelik (Azelik Mining Company), Niger
SOSUCAM	Société Sucrière du Cameroun (Cameroon Sugar Company)
SS	Schutzstaffel (Protection Squadron or Defence Corps), Nazi paramilitary organization
SYNPA	Association Synergie Paysanne (Farmers' Synergy Association), Benin
UNAIDS	Joint UN Programme on HIV/AIDS
UNCTAD	UN Conference on Trade and Development
UNDP	UN Development Programme
UNHRC	UN Human Rights Council (formerly UN Sub-Commission on the Promotion and Protection of Human Rights)
UNICEF	United Nations Children's Fund
UNOHCI	UN Office of the Humanitarian Coordinator in Iraq
UNRWA	United Nations Relief and Works Agency for Palestine Refugees in the Near East
VAM	Vulnerability Analysis and Mapping, WFP Food Security Analysis Unit
VAR-Palmares	Vanguardia Armada Revolucionaria–Palmares (Armed Revolutionary Vanguard–Palmares), Brazil
WARDA	West Africa Rice Development Association (now the Africa Rice Center)
WEF	World Economic Forum, Davos, Switzerland
WFP	UN World Food Programme
WHO	UN World Health Organization
WMO	UN World Meteorological Organization

PREFACE

The destruction, every year, of tens of millions of men, women, and children from hunger is the greatest scandal of our era. Every five seconds, a child under the age of ten dies of hunger—on a planet abounding in wealth and rich in natural resources. In its current state, the global agricultural system would in fact, without any difficulty, be capable of feeding 12 billion people, or twice the world's current population. Hunger is thus in no way inevitable. Every child who starves to death is murdered.

Relying upon the mass of statistics, graphs, reports, resolutions, and other studies released by the United Nations, organizations that specialize in problems of hunger, and other research institutions, as well as various NGOs, I attempt, in the first part of this book, to describe the extent of world hunger, and to assess the scope of the mass destruction it causes.

Almost half of the 56 million civilian and military deaths during World War II were caused by hunger and its immediate consequences. Like the biblical plague of locusts, the plundering Nazis descended upon the occupied countries, requisitioning the harvests, national food reserves, and livestock. For the inmates of the concentration camps, Hitler conceived, before the implementation of the Final Solution, the *Hungerplan* (hunger plan or hunger strategy), a program of planned starvation that aimed to liquidate as many detainees as possible through deliberate and prolonged deprivation of food.

Yet despite the European peoples' suffering, their collective experience of starvation had, in the immediate aftermath of the war, beneficial consequences. Several important researchers, patient prophets whom no one had heeded before, all at once saw hundreds of thousands of copies of their books sold and translated into a great many languages. The one universally recognized figure in this movement was Josué de Castro, a doctor born of mixed ethnic heritage in the impoverished northeastern provinces of Brazil, whose book *Geopolítica da fome* (The Geopolitics of Hunger), originally published in Portuguese in 1951, became known worldwide. Other writers, emerging in many different countries at about the same time, likewise exerted a profound influence on the collective consciousness—and the collective conscience—of the West, including Tibor Mende, René Dumont, and Abbé Pierre.

Immediately after its founding in June 1945, the United Nations created the Food and Agriculture Organization (FAO) and, not long after, the World Food Programme (WFP). In 1946, the UN launched its first global campaign against hunger. Finally, on December 10, 1948, the UN General Assembly, meeting in the Palais de Chaillot in Paris, adopted the Universal Declaration of Human Rights, whose article 25 defined the right to adequate nutrition. The second part of this book gives an account of this crucial moment in the awakening of the conscience of the West.

However, this moment was, unfortunately, very short-lived. Within the heart of the UN system itself, as well as in many of its member states, there were (and still are today) many powerful enemies of the right to food. The third part of this book unmasks them.

Deprived of adequate means to fight against hunger, the FAO and the WFP survive today only under highly adverse conditions. While the WFP succeeds, with great difficulty, in providing some of the emergency food aid needed by starving communities around the world, the FAO lies in ruins. The fourth part of this book reveals the reasons for the organization's decline.

In recent years, new scourges have descended upon the starving peoples of the southern hemisphere: expropriation of land by bio-fuel corporations and speculation in staple foods on commodities exchanges. The global power of the multinational corporations that dominate the agri-food industry and the hedge funds that speculate on the prices of agricultural commodities is superior to the power of national governments and all intergovernmental organizations. The leaders of agri-food and finance companies decide every day who on this planet will die and who will live.

The fifth and sixth parts of this book explain how and why, today, the obsession with profit, the lure of gain, the limitless greed of the predatory oligarchies of the globalized financial services industry prevail—both in public opinion and in governmental circles—over every other consideration, blocking effective action against hunger worldwide.

I was the first UN Special Rapporteur on the Right to Food. Together with my colleagues, men and women of exceptional abilities and commitment, I worked in this capacity for eight years. I would like especially to acknowledge Sally-Anne Way, Claire Mahon, Ioana Cismas, and Christophe Golay. Without these young scholars, nothing would have been possible. This book represents eight years of shared experiences and battles fought together.

I refer often throughout this book to the missions that we have undertaken in countries around the world stricken with famine: India, Niger, Bangladesh, Mongolia, Guatemala, and many others. Our reports from each mission reveal with particular clarity the devastation of the communities most severely afflicted by hunger. They reveal as well those who are responsible for this mass destruction. Doing so has not always been easy.

Mary Robinson is the former president of the Republic of Ireland and the former UN High Commissioner for Human Rights. Few of the bureaucrats at the UN could forgive this exceptionally elegant, keenly intelligent woman for her fierce sense of humor. In 2009, there were 9,923 international conferences, meetings of experts, and multilateral negotiation sessions among member states at the Palais des Nations, the European headquarters of many UN agencies in Geneva. There were even more in 2010. Many of these meetings concerned questions of human rights, and especially the right to adequate nutrition. Throughout her term of office, Mary Robinson showed little regard for most of these meetings. They smacked too much, she would say, of "choral singing"—referring to the old Irish tradition of village choirs that go from house to house on Christmas Day, singing in unison, year in and year out, the same trite songs. Indeed, there are hundreds of conventions in international law, intergovernmental organizations, and NGOs whose reason for being is to curb hunger and malnutrition. And in fact, from one continent to the other, thousands of diplomats, all year long, engage in such "choral singing" about human rights, while nothing ever changes in the lives of the victims of hunger. We must understand why.

How many times have I heard, during the debate that would follow my speeches in France, Germany, Italy, or Spain, such objections as, "But *monsieur*, if the Africans would only stop having children all over the place, they would be less hungry!" The ideas of Thomas Malthus die hard.

And what can one say of the lords of the corporate agri-food industry, the eminent directors of the World Trade Organization (WTO) and the International Monetary Fund (IMF), the "tiger shark" speculators, and the vultures who feed on the "green gold" of the biofuel industry, all of whom pretend that hunger is a natural phenomenon that can only be vanquished by Nature herself—that is, by a somehow self-regulating world market? According to them, such a market must, of course, inevitably create riches that would quite naturally benefit the hundreds of millions of starving people.

All consciousness is mediated. The world is not self-evident; it does not offer itself to view immediately as it really is, even to those who can see clearly. Ideology obscures reality. And crime, for its part, advances in disguise.

The older generation of Marxists of the Frankfurt School, such as Max Horkheimer, Ernst Bloch, Theodor Adorno, Herbert Marcuse, and Walter Benjamin, reflected at length on the individual's mediated perception of reality, and on the processes through which subjective consciousness is alienated by the doxa of an ever more aggressive and authoritarian capitalism. They sought to analyze the effects of the dominant capitalist ideology, especially the way in which that ideology leads people, from childhood, to consent to submit their lives to distant ends by depriving them of the possibilities of personal autonomy through which they might assert their freedom.

Some of these philosophers speak of a "double history": on the one hand, the visible history of everyday events, and on the other, the invisible history of consciousness. They show that consciousness is developed by hope in history, by a utopian spirit, by active faith in freedom. Such hope has a secular eschatological dimension: it nourishes an underground history that opposes to the actual justice system the justice that we deserve.

"It is not only the direct use of violence that has enabled the established order to maintain itself, but the fact that men themselves have learned to approve of it," writes Horkheimer. In order to change reality, to liberate the liberty latent within us, we must reawaken that "anticipatory consciousness," that historical force whose name is utopia, revolution.

Today, our awareness of the inevitability of progress is steadily growing. In the dominant Western societies above all, more and more women and men are mobilizing, fighting, confronting the neoliberal doxa that accepts the inevitability of mass starvation. More and more, it becomes irrefutable that hunger is human-made and that human beings can conquer it.

The question remains: how can we strike down this monster?

Deliberately ignoring Western public opinion, powerful revolutionary forces are awakening among the small farmers of the southern hemisphere. International farmers' unions, leagues of farmers who raise crops and livestock, are fighting against the vultures of "green gold" and against the speculators who seek to steal their land. They constitute the principal force in the battle against hunger.

In the epilogue to this book I return to this battle and the hope that nourishes it. Supporting it, for all of us, is a matter of life and death.

PART I

MASSACRE

1

THE GEOGRAPHY OF HUNGER

The human right to food, which follows from article 11 of the International Covenant on Economic, Social and Cultural Rights, has since 2002 been defined by the office of the UN's Special Rapporteur on the Right to Food as follows:

> The right to have regular, permanent and unrestricted access, either directly or by means of financial purchases, to quantitatively and qualitatively adequate and sufficient food corresponding to the cultural traditions of the people to which the consumer belongs, and which ensure a physical and mental, individual and collective, fulfilling and dignified life free of fear.

Among all human rights, the right to food is certainly the one most constantly violated on our planet. Allowing people to starve borders on organized crime. As we read in Ecclesiastes: "A meagre diet is the very life of the poor, to deprive them of it is to commit murder. To take away a fellow-man's livelihood is to kill him, to deprive an employee of his wages is to shed blood."

According to estimates made by the FAO, the number of people on the planet who are seriously and permanently undernourished reached 925 million in 2010, as against 1.023 billion in 2009.

Nearly a billion human beings out of the 7 billion on the planet thus suffer from permanent hunger.

The phenomenon of hunger may be approached in very simple terms. Solid foods, whether of animal or vegetable (and sometimes mineral) origin, are consumed by living beings to satisfy their needs for energy and nutrition. Liquid foods, or beverages (including water from underground sources, which may contain dissolved minerals), are consumed for the same purpose (liquid foods may be essentially considered solid food when they are in the form of soups, sauces, and so on). Together, solid and liquid sources of nourishment constitute what we call an organism's diet.

The human diet provides the vital energy that human beings need to live. The fundamental unit of food energy is the calorie, which enables us to measure the amount of nourishment that the body needs to grow, maintain, and rebuild itself. An inadequate caloric intake leads first to hunger, then to death. Human caloric needs vary according to age: about 700 calories per day for an infant, 1,000 for a child between one and two years old, and 1,600 for a five-year-old; adults' needs range from 2,000 to 2,700 calories per day depending on the climate where they live and the kind of work they do. The World Health Organization (WHO) sets 2,200 calories per day as the minimum necessary for an adult. Below this limit, an adult cannot maintain his or her body in a healthy state.

Severe, permanent undernutrition also causes acute suffering, tormenting the body. It induces lethargy and gradually weakens both mental and physical capacities. It leads to social marginalization, the loss of economic autonomy, and, of course, permanent unemployment on account of the sufferer's inability to engage in regular work. With rare exceptions, a human being may live normally for three minutes without breathing, three days without drinking, and three weeks without eating. No more. Then we begin to decline. Severe hunger leads inevitably to death.

To die of hunger is painful. The dying process is long and

causes unbearable suffering. Hunger destroys the body slowly, and it destroys the mind and spirit also. Anxiety, despair, a panicked feeling of being alone and abandoned accompany the body's physical decline.

Death from hunger passes through five stages. The body exhausts first its reserves of sugar, then of fat. Lethargy sets in, then rapid weight loss. Next the immune system collapses. Diarrhea accelerates the dying process. Oral parasites and respiratory tract infections cause dreadful suffering. Next the body begins to devour its own muscle mass. For undernourished children, death comes much more quickly than for adults. At the end, children can no longer stand upright. Like so many little animals, they huddle in the dust. Their arms hang lifelessly. Their faces look like those of the very old. Finally, they die.

In humans, neuronal development in the brain occurs primarily in the first five years of life. If, during this period, a child does not receive quantitatively and qualitatively adequate and sufficient food, his brain will remain stunted for life. By contrast, for example, an adult whose car breaks down while crossing the Sahara and who is deprived of food for some time before being saved, even at death's door, can return without difficulty to a normal life. A program of "re-nutrition" administered under medical supervision will enable a starving adult to regain all his or her mental and physical capacities.

The case of a child under five years of age deprived of sufficient food of adequate quality is entirely different. Even if such a child subsequently enjoy a series of miraculously favorable events in her life—her father finds work, she is adopted by a well-off family, and so on—her destiny is sealed. She has been crucified at birth; she will remain cognitively impaired for life. No therapeutic feeding program can provide her the satisfying, normal life she deserves.

In a great many cases, undernutrition causes illnesses called the "diseases of hunger": noma, kwashiorkor, and others. In addition, hunger dangerously weakens the immunological defenses of its victims. In his large-scale investigation of AIDS, Peter Piot, executive director of the Joint UN Programme on HIV/AIDS

(UNAIDS), has shown that millions of people who die of the disease could be saved, or could at least resist this scourge more effectively, if they had access to regular and sufficient nourishment. As Piot writes:

> For the poor across the globe, food is always the first necessity. Even more so in the face of HIV/AIDS. Good nutrition is the first line of defence in warding off the detrimental effects of the disease. And while it cannot match the effectiveness of life-extending drug therapies, nutritious food can help people infected with HIV stay healthier, longer. This allows teachers to continue to teach, farmers to continue to farm and parents to continue to care for their children. Without proper nutrition, however, the disease progresses faster and with more force.

In Switzerland, the average life expectancy at birth for men and women combined is slightly more than eighty-three years. In France, it is eighty-two. It is thirty-two years in Swaziland, a small country in southern Africa ravaged by AIDS and hunger.

The curse of hunger is passed from mother to child biologically. Every year, millions of undernourished women give birth to millions of children who are condemned from birth, deprived from their first day on earth. During her pregnancy, the malnourished mother transmits the curse of hunger to her child. Fetal undernutrition causes permanent physical and cognitive impairment: brain damage and neuromuscular motor deficiency. A starving mother cannot breast-feed her baby, nor does she have the means to buy infant formula. In the countries of the South, half a million women die in childbirth every year, most because of prolonged lack of food during pregnancy. Hunger is thus by far the leading cause of death and needless suffering on our planet.

How does the FAO attempt to collect data on world hunger? The organization's analysts, statisticians, and mathematicians are universally recognized for their expertise. The mathematical model that they developed first in 1971 and have been refining ever since is extremely complex. On a planet where 7 billion human beings live divided among some 193 states, it is obviously impossible to collect data on individuals. The FAO's statisticians therefore use an indirect method of sampling, which I describe in a deliberately simplified fashion here.

First, for each country the FAO gathers data on food production and on the country's imports and exports of foodstuffs, assessing for each of these figures the total number of calories represented. (Such an analysis reveals, for example, that even though India accounts for almost half of the people in the world who suffer from serious, permanent undernutrition, the country in certain years exports tens of millions of metric tons of wheat. Between June 2002 and November 2003, for example, India's wheat exports reached 17 million tons.) By this method, the FAO calculates the total number of calories available in each country.

Second, statisticians analyze for each country the population's demographic and sociological structure. As we have seen, caloric needs vary according to age. Sex constitutes another key variable: women burn fewer calories than men, for a whole range of sociological reasons. The work a person does and his socioeconomic status constitute still another important variable: a steelworker laboring at a blast furnace obviously requires more calories than a retiree who spends his days sitting on a park bench. Such factors vary furthermore according to the region and climatic zone under consideration; prevailing air temperatures and weather conditions influence caloric needs.

At this second stage of their analysis, FAO statisticians are in a position to correlate each country's caloric and demographic data, to determine its total caloric deficit, and therefore to calculate the theoretical number of people afflicted with serious, permanent undernutrition. However, the results of such calculations say nothing about the distribution of calories within a given population. The statisticians therefore refine their models by targeted surveys based on sampling techniques. The goal is to identify particularly vulnerable groups.

Bernard Maire and Francis Delpeuch have criticized the FAO's model. First, they question its *parameters*. The FAO's statisticians in Rome, they say, are able to determine nutritional deficits so far as calories are concerned, that is, at the level of macronutrients (protein, carbohydrates, fats) that provide calories, and therefore food energy. But they are utterly unable to account for a population's deficiencies in micronutrients, the lack of vitamins, minerals, and trace elements. Yet the absence in the food supply of enough iodine, iron, and vitamins A and C, among other elements indispensable to health, each year leaves millions of people blind, deformed, or disabled, and kills millions more. Thus the FAO manages with its statistical methods to calculate the number of victims of undernutrition, but not those who suffer from malnutrition.

Maire and Delpeuch further question the reliability of the FAO's *method*, which depends entirely upon the quality of the statistics provided to it by individual states. Many countries in the southern hemisphere, for example, have no system for gathering statistical information at all, not even in embryonic form. Yet it is precisely in the Southern countries that hunger claims the greatest number of victims.

Despite all the criticisms leveled at the mathematical model used by the FAO's statisticians—criticisms whose pertinence I recognize—I for my part consider that the model does enable us to grasp long-term variations in the number of undernourished people and deaths from hunger on our planet. In any case, even if the FAO's figures are underestimates, its method does satisfy this dictum of Jean-Paul Sartre: "To know the enemy is to fight the enemy."

The current goal of the UN is, by 2015, to reduce by half the number of people suffering from hunger. In formally adopting this target in 2000 as the first of the UN's eight Millennium Development Goals (MDGs), the Assembly General in New York took 1990 as its point of reference. It is thus the total number of people who were starving in 1990 that the UN is attempting to reduce by half.

This goal, of course, will not be reached, for the pyramid of martyrs to hunger, far from shrinking, only grows. As the FAO itself admits:

> Latest available statistics indicate that some progress has been made towards achieving MDG 1, with the prevalence of hunger declining from 20 percent undernourished in 1990–92 to 16 percent in 2010. However, with the world's population still increasing (albeit more slowly than in recent decades), a declining proportion of people who are hungry can mask an increase in the number. In fact, developing countries as a group have seen an overall setback in terms of the number of hungry people (from 827 million in 1990–92 to 906 million in 2010).

In order to more accurately determine the geography of hunger, the distribution of this form of mass destruction around the planet, we must first have recourse to a fundamental distinction that is referred to by the UN and its specialized agencies: "structural hunger" on the one hand, and "conjunctural hunger" on the other.

Structural hunger inheres in the insufficiently developed structures of agricultural production in the South. It is permanent and unspectacular, and it is reproduced biologically as, every year, millions of undernourished mothers bring millions of hungry babies into the world. Structural hunger represents the physical and mental destruction of human beings, the shattering of their dignity, endless suffering.

Conjunctural hunger, on the other hand, is highly visible. It erupts periodically on our television screens. It occurs suddenly, when natural disasters—swarms of locusts, drought, floods—devastate a region, or when war tears apart the fabric of a society, ruins its economy, drives hundreds of thousands of its victims into camps for internally displaced persons within a country or into refugee camps beyond its borders. In all these situations, farmers can neither sow seed nor harvest their crops anymore. Markets are destroyed, roads blocked, bridges collapsed. The institutions of government no longer function. For millions of victims of hunger penned into camps, the WFP is their only hope.

Nyala, in Darfur, is the largest of the seventeen camps for internally displaced persons in the three provinces in western Sudan ravaged by civil war and hunger. Guarded by African "blue helmets" (UN peacekeeping forces), mostly Rwandese and Nigerian, nearly a hundred thousand undernourished men, women, and children are crammed together in an immense camp under canvas and plastic sheeting. Any woman who ventures out even five hundred meters from the camp fence—to search for firewood or well water—runs the risk of being captured by the Janjawid, the mounted Arab gunmen hired by the Islamist dictatorship in Khartoum, the capital of Sudan. She will certainly be raped, and possibly murdered.

If the WFP's trucks, white Toyotas topped by blue UN flags, do not arrive every three days with their pyramidal loads of sacks of rice and flour, containers of water, and crates of medicine, the Zagawha, Massalit, and Fur people confined behind barbed wire and protected by the blue helmets will soon perish.

Who are the people at greatest risk of hunger? The three most vulnerable large groups are, in the FAO's terminology, the rural poor, the urban poor, and victims of natural and human-made disasters described above. Let us pause to consider the first two of these categories.

THE RURAL POOR

The majority of human beings who do not have enough to eat belong to communities of the rural poor in the South. Many have access to neither potable water nor electricity. In these areas, services that provide public sanitation, education, and hygiene are for the most part nonexistent. Of the 7 billion human beings on the planet, slightly less than half live in rural areas.

Since the dawn of time, peasant communities—farmers and pastoralists (and fishers as well)—have always been among the first victims of extreme poverty and hunger. Today, of the 1.2 billion human beings who, according to World Bank criteria, live in "extreme poverty" (that is, on an income of less than $1.25 per day), 75 percent live in the countryside.

Many agricultural workers live in extreme poverty for one or another of the following three reasons. Some are landless migrant workers or tenant farmers overexploited by landowners. Thus, for example, in northern Bangladesh, Muslim tenant farmers are forced to remit to their Hindu landlords who live in Calcutta four-fifths of their harvest. Others, if they do have land, do not enjoy sufficiently secure title to it. This is the case of the Brazilian *posseiros*, who occupy small areas of unproductive or vacant land, which they work without holding documents proving that the land belongs to them. For still others, even if they have clear title to their land, their fields are insufficient in extent and quality to feed their families decently.

The International Fund for Agricultural Development (IFAD) estimates the number of landless agricultural workers at around 500 million, representing some 100 million households. These are the poorest of the poor on earth.

For small farmers, overexploited tenant farmers, agricultural day laborers, and migrant workers, the World Bank henceforward recommends "market-assisted land reform," which it first advocated in 1997 for the Philippines. According to this scheme, the landowning classes would be obliged to give up part of their holdings, but rural workers would have to buy the land, possibly with the help of credits from the World Bank. In view of the utter destitution of landless families, "market-assisted" agrarian reform, promoted worldwide by the World Bank, amounts to the most blatant hypocrisy, not to say indecency pure and simple.

The liberation of poor farmers can only be accomplished

by the farmers themselves. Anyone who has ever attended an *assentamento* (settlement) or an *acampamento* (encampment) of the Movimento dos Trabalhadores Rurais Sem Terra (MST; Landless Rural Workers' Movement) of Brazil feels moved and filled with admiration. The MST has become the most important social movement in Brazil, dedicated to agrarian reform, food sovereignty, questioning the assumptions of free trade and the dominant corporate food industry model of food production and consumption, promotion of subsistence agriculture, solidarity, and internationalism.

The international rural workers' movement known as La Via Campesina brings together 200 million tenant farmers, smallholders (who own 1 hectare—about 2.5 acres—or less), and seasonal rural workers across the world, including both sedentary and nomadic pastoralists, as well as self-employed fishers. The organization's headquarters is in Jakarta, Indonesia. La Via Campesina is today one of the most impressive revolutionary movements in the Third World. I will discuss its work further in later chapters.

Few men and women on earth work as hard, in such adverse climatic conditions, and for such meager return on their efforts as the small farmers of the southern hemisphere. Few among them are able to save anything from their earnings to protect themselves against the natural disasters, locusts, and social unrest that always threaten them. Even if, for a few months at a time, food is available in abundance—as drums sound and marriages are celebrated with sumptuous feasting, and everyone shares whatever they have—the threat of hunger is omnipresent. No one can ever know for certain how long they will have to wait, after exhausting the last year's harvest, to gather the new harvest in—a "hunger gap" during which food prices often increase substantially, and during which the rural poor must either buy food or starve.

Ninety percent of small farmers in the South have no tools other than the hoe, the machete, and the scythe. More than a billion have neither a draft animal nor a tractor. When a farmer doubles his pulling power, the area of land he can cultivate doubles likewise. Without animal or mechanical traction, the farmers of the South will remain confined to extreme poverty.

In the Sahel, 1 hectare planted with cereal crops yields between 600 and 700 kilograms (536 to 625 pounds per acre). In Brittany, Beauce, Baden-Wurtemberg, or Lombardy, 1 hectare of wheat yields 10 tons—more than ten times as much. This difference in productivity cannot of course be explained by a disparity in skills. The Bambara, Wolof, Mossi, and Toucouleur farmers of West Africa work with the same energy, the same intelligence as their European colleagues. The difference lies in the inputs available to each. In Benin, Burkina Faso, Niger, or Mali, most farmers do not enjoy the advantage of any irrigation system; like farmers three thousand years ago, they rely on the rains—only 3.8 percent of farmland in sub-Saharan Africa is irrigated (versus 37 percent in Asia, for example). Moreover, they also have at their disposal neither mineral fertilizers, nor improved seed, nor pesticides to fight predatory insects.

Indeed, the FAO estimates that 500 million farmers in the South as a whole have no access to improved seed, mineral fertilizers, or even manure (or other natural fertilizers), because they have no animals. Again according to the FAO, 25 percent of the world's harvests are destroyed each year by bad weather or rodents. Silos are rare in sub-Saharan Africa, in South Asia, and on the Altiplano (the Andean plateau). It is thus the farming families of the South who are the first and hardest hit by harvest destruction.

The transportation of harvested crops to market is another great problem. Farmers who cannot keep their rice dry can store it for only four months before it begins to germinate. If it does not get to market within this period, it is lost. I witnessed in Ethiopia in 2003 the following absurd scenario: in Mek'ele, in the Tigray region, in the high plateaus tormented by winds, where the earth is cracked and dusty, famine ravaged 7 million people. Yet 500 kilometers farther west, in Gondar, tens of millions of tons of teff rotted in the granaries for lack of trucks and roads capable of transporting food that could have saved countless lives.

In sub-Saharan Africa, in India, or in the heart of the Aymara and Otavalo communities of the Altiplano in Peru, Bolivia, and Ecuador, there is no such thing as a cooperative rural retail bank on the model of the French and Swiss *crédit agricole*. As a result, farmers have no choice: most of the time, they have to sell their crops at the worst possible moment, that is, immediately after harvest, when prices are lowest.

Once they are caught in the spiral of overindebtedness—descending ever deeper into debt in order to pay the interest on their previous debts—farmers are forced to sell their future harvest to be able to buy, at a price fixed by the lords of the global agrifood trade, the food their families need to bridge the hunger gap.

In rural areas, especially in Central and South America, India, Pakistan, and Bangladesh, violence is endemic.

Consider Guatemala. Together with my colleagues, I undertook a mission to that country from January 26 to February 5, 2005. During our stay, the Guatemalan government's commissioner for human rights, Frank La Rue, himself a former leader of the opposition to the dictatorship of General Efraín Ríos Montt, brought to my attention the crimes committed day after day against the small farmers of his country.

On January 23, in the *finca* (rural plantation or agricultural estate) of Alabama Grande, an agricultural worker stole some fruit. Three of the *finca*'s security guards found and killed him. That evening, when their husband and father did not return home, his family, who, like all the families of the *péones* (landless rural laborers), lived in a hut on the edge of the estate, grew worried. Accompanied by their neighbors, the man's eldest son, who was fourteen years old, went up to the estate owner's house. The guards intercepted them. A dispute broke out. Voices were raised. The guards beat the boy and four of the men who were with him. On another *finca*, other guards stopped a young boy whose pockets were full of *cozales*, a local fruit. Accusing him of having stolen the fruit from the landowner's fields, they handed him over to the *patrón*, who killed the boy with a pistol shot.

La Rue said to me, "Yesterday, in the presidential palace, the vice president of the republic, Eduardo Stein Barillas, explained it to you like this: forty-nine percent of the children under ten years old are undernourished . . . ninety-two thousand of them died of hunger and diseases caused by hunger last year. . . . So you can understand why fathers, brothers, at night . . . they climb into the *finca*'s orchards . . . they steal some fruit, a few vegetables. . . ."

In 2005, 4,793 murders were committed in Guatemala, 387 in the course of our brief stay. Among the victims were four young members of a farmers' union, three men and a woman, who had just returned from a training course in Fribourg, Switzerland. Their killers machine-gunned their car in the Sierra de Chuacús mountains, on the road between San Cristóbal Verapaz and Salamá.

I heard the news during a dinner at the Swiss embassy. The ambassador, a determined man who loved and knew Guatemala very well, promised me that he would lodge a strong protest the very next day with the minister of foreign affairs. Also in attendance at this dinner was Rigoberta Menchú, winner of the Nobel Peace Prize in 1992, a magnificent Mayan woman who, under the dictatorship of General Fernando Romeo Lucas García, lost her own father and one of her brothers, who were burned alive. As we left, on the doorstep, she whispered to me, "I was watching your ambassador. He was pale . . . his hand was trembling. . . . He is furious. He's a good man. He will protest—but it will do no good!"

Near Finca las Delicias, a coffee estate in the *municipio* of El Tumbador, I questioned striking *péones* and their wives. For six months, the landowner had not paid his workers, using as his excuse the crash in coffee prices on the world market. (In 2005, the legal minimum wage in Guatemala was 38 quetzals a week, or about $5.) A demonstration organized by the striking workers had just been violently put down by the police and the landowner's private security guards.

Monsignor Álvaro Ramazzini Imeri, the bishop of the San Marcos diocese, president of the Conference of Catholic Bishops of Guatemala, and an important advocate for the poor and marginalized of Guatemala, had warned me: "Often, at night, after a demonstration, the police return and randomly arrest young people . . . who often disappear."

We are sitting on a wooden bench in front of a shack. The strikers and their wives are standing in a semicircle. In the night's muggy heat, their children watch us with serious expressions. The women and young girls wear vividly colored dresses. A dog barks in the distance. The sky is spangled with stars. The scent of the coffee trees mingles with the smell of the red geraniums growing behind the house. Clearly, these people are afraid. Their faces betray their anxiety, which has surely been fed by the nighttime arrests and the disappearances organized by the police that Bishop Ramazzini told me about. In a frankly clumsy fashion, I hand out my UN business cards. The women press them to their hearts, like talismans.

In the very moment that I am telling them about human rights and the possibility of UN protection, I know already that I am betraying them. The UN, of course, will do nothing. Hiding away in their villas in Guatemala City, the capital, the local UN officials content themselves with administering expensive so-called development programs—which profit the landowners. Maybe, nonetheless, Vice President Eduardo Stein Barillas, a former Jesuit and close to La Rue, will warn El Tumbador's chief of police about the possible "disappearances" that threaten the young strikers. . . .

The greatest violence done to the world's poor farmers is of course the unequal distribution of land. In Guatemala, in 2011, 1.86 percent of the population owned 57 percent of the arable land. There are thus in this country forty-seven large properties each extending over 3,700 hectares (9,143 acres) or more, while 90 percent of the farmers live on plots of 1 hectare or less.

As for the violent crimes committed against the farmers' unions and the protesting strikers, their situation has not improved. On the contrary: the number of disappearances and murders has increased.

THE URBAN POOR

In the world's urban shantytowns—the *callampas* of Lima, the slums of Karachi, the *favelas* of São Paulo, the squatter encampment at Manila's Smokey Mountain landfill—the mothers of poor families must, in order to buy their food, manage on an extremely limited budget. As I have said, the World Bank estimates that 1.2 billion "extremely poor" people live on less than $1.25 per day. In Paris, Geneva, or Frankfurt, a housewife spends on average 10 to 15 percent of her family income to buy food. In the budget of a woman living at Smokey Mountain, food accounts for 80 to 85 percent of her total expenses.

In Latin America, according to the World Bank, 41 percent of the population lives in "informal housing." The slightest increase in the prices of food in the shantytowns' local markets causes anxiety, hunger, disintegration of the family—catastrophe.

The separation of urban from rural poor is not of course as absolute as it first appears because in reality, as I have said, 43 percent of the world's 2.7 billion seasonal workers, smallholders, and tenant farmers, who constitute the majority of the poorest people living in rural areas, must also, at certain times of the year, buy food in the market of their local village or market town to survive the hunger gap between harvests. The rural worker thus bears the full brunt of rising prices for the food that he absolutely must procure.

Yolanda Areas Blas, Nicaragua's warm and energetic representative to La Via Campesina, offers the following example. The Nicaraguan government annually defines a *canasta básica*, a "basket" of consumer goods including twenty-four essential foodstuffs that a family of six needs each month to survive. In March 2011, the cost of the canasta básica in Nicaragua was 6,250 cordobas (about $500). The legal minimum wage in the same period for an agricultural worker (which is, moreover, rarely paid) was 1,800 cordobas (about $80).

The geographical distribution of world hunger is extremely unequal. In 2010, the situation looked like this:

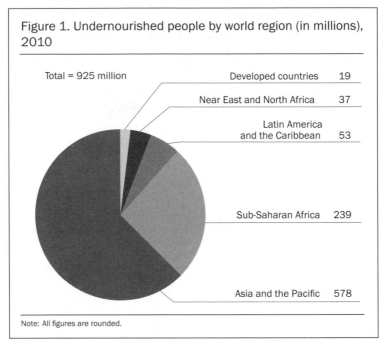

Figure 1. Undernourished people by world region (in millions), 2010

Total = 925 million

Developed countries	19
Near East and North Africa	37
Latin America and the Caribbean	53
Sub-Saharan Africa	239
Asia and the Pacific	578

Note: All figures are rounded.

The following table enables us to assess the variation over time of the total number of victims of hunger in recent decades:

Figure 2. Undernourished people worldwide, 1969–2007

Year	Number (in millions)	Percentage of total population
2005–7	848	13%
2000–2002	833	14%
1995–97	788	14%
1990–92	843	16%
1979–81	853	21%
1969–71	878	26%

The following table shows the change in the scale of hunger in the various regions of the world from 1990 to 2007, that is, over the course of approximately one generation:

Figure 3. Undernourished people by world region (in millions), 1990–2007

Region	1990–92	1995–97	2000–2002	2005–7
World	843.4	787.5	833	847.5
Developed countries	16.7	19.4	17	12.3
Developing countries	826.6	768.1	816	835.2
Asia & the Pacific (inc. Oceania)	587.9	498.1	531.8	554.5
East Asia	215.6	149.8	142.2	139.5
Southeast Asia	105.4	85.7	88.9	76.1
South Asia	255.4	252.8	287.5	331.1
Central Asia	4.2	4.9	10.1	6
Western Asia	6.7	4.3	2.3	1.1
Latin America & the Caribbean	54.3	53.3	50.7	47.1
North & Central America	9.4	10.4	9.5	9.7
The Caribbean	7.6	8.8	7.3	8.1
South America	37.3	34.1	33.8	29.2
Near East & North Africa	19.6	29.5	31.8	32.4
Near East	14.6	24.1	26.2	26.3
North Africa	5	5.4	5.6	6.1
Sub-Saharan Africa	164.9	187.2	201.7	201.2
Central Africa	20.4	37.2	47	51.8
East Africa	76.2	84.7	85.6	86.9
Southern Africa	30.6	33.3	35.3	33.9
West Africa	37.6	32	33.7	28.5
Africa (all regions)	169.8	192.6	207.3	207.2

These statistics, which end in 2007, should be compared to the changes in the world's population by continent, for which the figures in 2007 were as follows:

Figure 4. World population distribution by world region, 2007

Region	Population (millions)	Percentage of world population
Asia	4,030	60.5%
Africa	965	14.0%
Europe	731	11.3%
South America (inc. Central America & the Caribbean)	572	8.6%
North America	339	5.1%
Oceania (inc. Australia and New Zealand)	34	0.5%

The following graph shows long-term change in the scale of world hunger from 1969 to 2010, or over the course of about two generations:

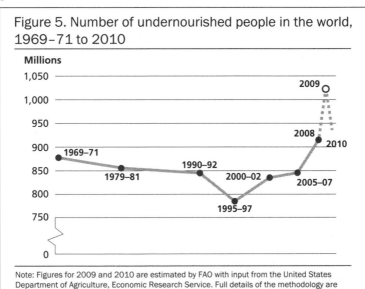

Figure 5. Number of undernourished people in the world, 1969–71 to 2010

Note: Figures for 2009 and 2010 are estimated by FAO with input from the United States Department of Agriculture, Economic Research Service. Full details of the methodology are provided in the technical background notes (available at www.fao.org/publication/sofi/en/).

This graph requires several qualifications. Obviously, we must compare these figures to the total global population increase over the course of the same period: in 1970, there were 3.696 billion people on the planet; in 1980, 4.442 billion; in 1990, 5.279 billion; in 2000, 6.085 billion; and in 2010, 6.7 billion. Since 2005, the global total number of victims of hunger has increased catastrophically, while the rate of increase of the world's total population—about 400 million people every five years—has remained stable. The greatest increase in the number of undernourished people occurred between 2006 and 2009, even though, according to the FAO's figures, harvests of cereal crops were good worldwide in this period. The number of undernourished people rose severely because of an explosion in the prices of food and because of the crisis that I analyze in the sixth section of this book, the scandal of market speculation in food.

The following graph offers a more detailed portrait of regional trends in hunger in the developing world from 1990 to 2010:

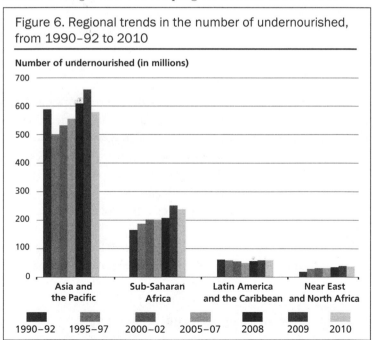

Figure 6. Regional trends in the number of undernourished, from 1990–92 to 2010

Number of undernourished (in millions)

This graph reveals that the developing countries in recent years have been home to between 98 and 99 percent of the undernourished people on the planet. In absolute terms, the region afflicted with the greatest number of starving people remains Asia and the Pacific, but with a reduction of 12 percent in recent years (from 658 million in 2009 to 578 million in 2010), this region showed genuine improvement by 2010. It is in sub-Saharan Africa that the percentage of undernourished people remains the highest at present, at 30 percent in 2010, or nearly one person in three.

If the majority of the victims of hunger live in developing countries, the industrialized West does not entirely escape the specter of hunger. Nine million seriously and permanently undernourished people live in the industrialized countries, and another 25 million in the nations of Eastern Europe and the former Soviet Union said to be "in transition."

In the first few months of 2011, once again, as in 2008, the world market in staple food commodities exploded. (The staple foods are rice, wheat, and corn [maize], which together account for 75 percent of world food consumption. Rice alone constitutes 50 percent of the world's total volume of food.) In February 2011, the FAO warned that eighty countries stood on the brink of food insecurity.

On December 17, 2010, the people of Tunisia rose up against the predators in power in Carthage (today a wealthy residential suburb of the capital city, Tunis, and the location of the presidential palace). President Zine El-Abidine Ben Ali, who, together with his in-laws and accomplices, had terrorized and plundered Tunisia for twenty-three years, fled to Saudi Arabia on January 14, 2011. The effect of the Tunisian uprising on neighboring countries was immediate.

In Egypt, the revolution began on January 25, with the gathering of nearly a million people in Tahrir Square, in the heart of Cairo. Since October 1981, Hosni Mubarak, a former officer in the Egyptian air force, had ruled the Israeli-American protectorate of Egypt by means of torture, police terror, and corruption. During the three weeks preceding his fall from power, elite secret-police snipers, perched on the roofs overlooking the square, murdered more than three hundred men and women, mostly young people, while the security forces "disappeared" hundreds of others into their torture chambers. Mubarak was overthrown on February 12. The unrest soon spread throughout the entire Arab world, including both North Africa and the entire Middle East: from Libya, to Yemen, to Syria, to Bahrain, and elsewhere.

The revolutions in Egypt and Tunisia have complex causes; the tremendous courage of the protesters in each country draws on deep roots. However, hunger, undernutrition, and anxiety provoked by the rapidly rising cost of their daily bread constitute a powerful source of their rebellion. Ever since the period when Tunisia was a French protectorate, the baguette has been the staple of the Tunisian diet, just as the *eish masri* or *eish balad*, a thick pita-type bread, has been the Egyptians' essential food. In June 2010, the price of a ton of bread wheat on the world market began to rise steeply, spiking in February 2011 at over $348, or more than double the price of June 2010.

The vast area that stretches from the Atlantic coast of Morocco to the emirates of the Persian Gulf is the world's largest importer of cereals. Whether in terms of cereals, sugar, feed for cattle and poultry, or oilseed products, all the countries of North Africa and the Gulf are massive importers of food. To feed its 84 million inhabitants, Egypt imports more than 10 million tons of wheat per year; Algeria, 5 million; Iran, 6 million; Morocco and Iraq between 3 and 4 million each. Saudi Arabia buys about 7 million tons of barley on the world market annually.

In Egypt and in Tunisia, the threat of famine had tremendous consequences, as the specter of hunger mobilized unprecedented forces that have led to the blossoming of the "Arab Spring." But in most of the other countries threatened by imminent food insecurity, the people continue to endure their suffering and fear in silence.

We must moreover understand that in the rural areas of Asia and Africa, women are subject to permanent discrimination that results in undernutrition. Thus among certain culture groups in

Sudan, Somalia, and the Sahel, women and their daughters eat only what is left after the men in the family and their sons have eaten. Very young children suffer from the same kind of discrimination. Widows and second or third wives must endure even more seriously discriminatory treatment.

In the Somali refugee camps in Kenya, the representatives of the UN High Commissioner for Refugees fight a daily battle against the following appalling custom. Among Somali pastoralists, women and young girls are not allowed to touch the bowl of millet or the remains of grilled mutton until after the men have had their meal. The men serve themselves, then let the male children have their turn. At the end of the meal, when the men have left the room with their sons, the women and girls gather round the bowls on the floor mat, in which the men have left behind a few little clumps of rice, leftover bits of wheat, a few shreds of meat. If the bowls are empty, the women and girls do not eat.

One more word about the victims of hunger. The geography that I have sketched and the statistics that I have listed account for at least one human being in seven on earth. But when we look at the situation from another point of view, when we cease to consider the death of a child as a mere statistic, but see it as the disappearance of a singular, irreplaceable being who comes into the world to live a unique life and who will never return, then the perennial destruction wrought by hunger—in a world abounding in riches and in which we seem capable of accomplishing so many "impossible" things—appears ever more unacceptable. It amounts to nothing less than a massacre, mass murder of the poorest of the poor.

2

INVISIBLE HUNGER

Alongside the human beings destroyed by undernutrition, the victims encompassed within the terrifying geography of hunger mapped in the previous chapter, there are also those whose lives are ravaged by malnutrition. The FAO does not neglect them, but counts them separately. While undernutrition is the result of a lack of calories, malnutrition is caused by deficiencies in micronutrients—vitamins and minerals. Several million children under age ten die of acute, severe malnutrition every year.

During my eight-year term as UN Special Rapporteur on the Right to Food, I traveled throughout the lands of hunger. On the arid, frozen heights of the Yucatán sierra in Guatemala, in the desolate steppes of Mongolia, deep in the dense jungle of Orissa in India, in villages stricken by the famine endemic to Ethiopia and Niger, I have seen gray, toothless women who, at age thirty, looked eighty years old; little boys and girls with huge black eyes, astonished, cheerful, but with arms and legs as thin as matchsticks; men, ashamed, gesturing slowly, with emaciated bodies. Their affliction is immediately visible. All are victims of a lack of sufficient calories.

The ravages of malnutrition, on the other hand, are not immediately visible. A man, woman, or child may be at normal weight and suffer nonetheless from malnutrition, that is, from serious, permanent deficiencies of vitamins and minerals essential to the assimilation of *macro*nutrients. These substances are called *micro*nutrients because they are necessary only in tiny quantities; nonetheless, the body cannot grow, mature, and maintain itself in good health without these nutrients, because the organism cannot synthesize them. They can only be provided by a diet that is varied, balanced, and of high quality.

Deficiencies in vitamins and minerals can in fact lead to very serious health problems: a greatly increased vulnerability to infectious diseases, blindness, anemia, lethargy, reduced learning ability, cognitive impairment, congenital deformities, and death. The three most common deficiencies are lack of vitamin A, iron, and iodine.

The UN often calls malnutrition "silent hunger." Yet victims of malnutrition may in fact cry out in pain. I prefer to speak of *invisible* hunger, imperceptible to the eye—even, oftentimes, to the eye of a doctor. A child's body may appear to be well fed, with normal curves, and she may weigh within the normal range for her age, while nonetheless malnutrition gnaws away at her from inside—a dangerous state that, as much as a sheer lack of calories, may lead to serious illness and death.

However, as critics have been pointed out, deaths due to malnutrition are not captured in the FAO's hunger statistics, which only count total available calories. Insofar as children under fifteen years old are concerned, UNICEF (the United Nations Children's Fund) and the not-for-profit Micronutrient Initiative (MI) have since 2004 undertaken periodic surveys whose results are published in a series of reports with titles such as *Vitamin and Mineral Deficiency: A Global Damage Assessment Report*. These reports show that one-third of the world's people fail to realize their full physical and intellectual potential because of vitamin and mineral deficiencies. Malnutrition is particularly devastating for children under five.

Anemia, due to iron deficiency, is one of the most frequent consequences of malnutrition. It is characterized chiefly by a low hemoglobin count in the blood. Anemia can be fatal, especially for children and women of childbearing age. For infants, iron is essential: the major part of neuronal development occurs within the first two years of life. Furthermore, anemia also causes immune system malfunction. About 30 percent of births occur in the world's fifty poorest countries or LDCs (least-developed countries), to use the UN's terminology. Among these babies, the lack of iron causes irreparable damage. Many will be cognitively impaired for life.

Every four minutes a human being goes blind, in most cases because of nutritional deficiency. Blindness is most often caused by a lack of vitamin A: 40 million of the world's children suffer from a lack of this nutrient; 13 million of them go blind every year. Beriberi, a disease that destroys the nervous system, is caused by prolonged lack of thiamine, or vitamin B1. The absence of vitamin C in food causes scurvy and, in small children, rickets. Folic acid, also called vitamin B9, is absolutely essential to pregnant women; WHO estimates that 200,000 babies are born every year suffering from birth defects due to a lack of this micronutrient.

Iodine is essential to good health. Nearly a billion people— mainly men, women, and children living in the southern hemisphere, especially those who live in mountainous regions and in floodplains where both drinking water and the water-saturated soil are often low in iodine—suffer from a natural deficiency of iodine in their diet. When this deficiency is not compensated for by supplementation, it causes goiter in adults (enlargement of the thyroid gland, which leads to severe swelling of the neck) and cretinism in children, a syndrome of severely stunted physical and mental growth caused by untreated congenital deficiency of thyroid hormones (congenital hypothyroidism) due to maternal hypothyroidism. For a pregnant mother and consequently also for the fetus, a lack of iodine can be fatal.

A lack of zinc also affects both motor and mental abilities. According to a study undertaken by *The Economist*, zinc deficiency causes some 400,000 deaths annually. Zinc deficiency also causes diarrhea, often fatal, in young children.

It is important as well to realize that half the people afflicted with a lack of micronutrient suffer from cumulative deficiencies, meaning that they suffer from the lack of a sufficient amount of several vitamins and several minerals at the same time.

Fully half of the deaths of children under age five around the world are caused directly or indirectly by malnutrition. The great majority of these children live in South Asia or sub-Saharan Africa. Only a small percentage of the world's malnourished children have access to treatment: national health policies in numerous countries in the South only rarely address acute, severe malnutrition, even though the condition can be treated at minimal cost and treatment does not pose any particular therapeutic problems. Health centers specializing in the treatment of malnutrition are sorely lacking in many countries. In a document released in 2008, Action Contre la Faim, the worldwide NGO known in English as Action Against Hunger, rightly protested: "To end childhood malnutrition would be easy. We have to make it a priority. Yet many governments lack the will."

In all probability, since 2008, the situation has if anything grown worse. In sub-Saharan Africa, for example, primary sanitation services have continued to deteriorate. In Bangladesh, where the number of malnourished children exceeds 400,000, there are only two hospitals capable of administering the care that can bring back to life a little boy or girl ravaged by a lack of vitamins and/or minerals.

We must, moreover, not forget that malnutrition, like undernutrition, has severe psychological as well as physical consequences. The lack of macro- and micronutrients and the host of illnesses such deprivation entails cause anxiety, a permanent sense of shame, depression, an obsessive fear of each new day. How will a mother whose children cry from hunger at night, and who miraculously succeeds in borrowing a little bit of milk from her neighbor, be able to feed them tomorrow? How can she avoid going mad? What father, unable to feed his family, can fail to lose, in his own eyes, every last shred of his dignity? A family denied regular access to food, of adequate quality and in sufficient amounts, is a family destroyed. The tens of thousands of farmers in India who have committed suicide bear tragic witness to this harsh truth.

3

PROTRACTED CRISES

A t the center of the FAO's analysis of world hunger is the concept of protracted crisis—a rather banal term used by many UN agencies that fails to do justice to the tragedies, contradictions, tensions, and failures that mark the world's most long-lasting food crises. During a protracted crisis, structural hunger and conjunctural hunger mutually reinforce each other. A natural disaster, a war, or a plague of locusts destroys the economy, causing society to disintegrate and weakening institutions of all kinds. A country thus afflicted can no longer find a way out of the crisis or recover a minimal sense of balance. The state of emergency becomes an ordinary way of life for the inhabitants. Tens, even hundreds of millions of people thrown into this situation struggle in vain to reconstruct a society destroyed by hunger. But food insecurity is only the most obvious external symptom of protracted crisis. Such crises are by no means all the same, but they do share certain characteristics:

Long duration. Afghanistan, Somalia, and Sudan are all examples of countries that have been in a crisis situation since the 1980s—for three decades.

Armed conflicts. War may afflict only one relatively isolated region of a country, as in Uganda, in Niger, or in Sri Lanka from 2000 to 2009. Or, on the contrary, it may engulf an entire

country, as was until recently the case in Liberia and Sierra
Leone.

The weakening of institutions. In protracted crises, governmental
and other public institutions are weakened to an extreme degree,
whether because of corruption among leaders or lower-level officials, or owing to the disintegration of the social fabric caused
by war.

All the countries in a state of protracted crisis are included
on the list of the fifty least developed countries. This list of the
LDCs is compiled annually by the UN Development Programme
(UNDP) according to the following criteria: low income (based
on a three-year average estimate of the gross national income
per capita); human resource weakness (based on indicators of
nutrition, health, education, and adult literacy); and economic
vulnerability (based on indicators of such factors as instability of
agricultural production and exports, degree of industrialization,
reliance on exports, and percentage of the population displaced
by natural disasters).

Currently twenty-one countries fulfill the FAO's criteria for
protracted crisis. All these countries have experienced a human-
caused emergency—military conflict or political crisis. Eighteen
of them have also had to confront, at one time or another, a natural disaster, in isolation or in combination with a military/politi-
cal crisis.

Niger is a magnificent country in the Sahel extending over
nearly 1.27 million square kilometers (about half a million square
miles), home to some of humanity's most splendid cultures, including those of the Djerma, Hausa, Touareg, and Fula peoples.
Niger is also a typical example of a country in a protracted state of
crisis. Niger has very little arable land: only 4 percent of the land
is entirely suitable for agriculture. Aside from the Djerma and
some of the Hausa, most of the country's peoples are descended
from nomadic or seminomadic pastoralists. Niger has 20 million
head of cattle, white camels, zebus (humpbacked cattle with lyre-
shaped horns), goats (notably the pretty red Maradi goat), sheep,
and donkeys. In the center of the country, the soil is rich in min-

eral salts that give the livestock who lick them extraordinarily firm and tasty flesh.

But the people of Niger are crushed under the burden of their external debt, and are subjected on that account to the iron law of the International Monetary Fund (IMF). In the course of the last ten years, the IMF has ravaged the country with a series of "structural adjustment" programs. In particular, the IMF ordered the closing of the National Veterinary Office, opening the market to multinational veterinary pharmaceutical corporations. The government of Niger now no longer has any control over the expiration dates of veterinary vaccines and medications. (Niger's capital city, Niamey, is 1,000 kilometers [620 miles] from the Atlantic coast. Many veterinary medicines have expired by the time they reach Niamey's markets. Local merchants make do by changing the expiration dates on the labels by hand.) Since the closing of the Veterinary Office, the herders of Niger have been forced to buy antiparasitic medicines, vaccines, and other drugs and supplements to treat their herds at prices dictated by Western multinational corporations.

The climate in Niger is harsh. To keep a herd numbering several hundred or several thousand head healthy is expensive. Most herders are in no way able to pay the new prices. As a result, their animals fall sick and die. At best, they can be sold for next to nothing before they die. Human health, which is directly linked to the health of the animals, deteriorates. The proud herdsmen sink into despair and tumble down the social ladder. Together with their families, they migrate to the shantytowns of Niamey or Kano, Niger's second-largest city after Lagos, in the north, or the big coastal cities of neighboring countries, such as Cotonou, Abidjan, or Lomé.

In this country of recurrent famine, where drought periodically exposes people and their animals to undernutrition and malnutrition, the IMF has forced the government to dismantle its system of food reserves, which amounted to 40,000 tons of cereals. The state used to store mountains of sacks of millet, barley, and wheat in its warehouses precisely in order to be able, on an emergency basis, to help the country's most vulnerable communities survive drought, invasions of locusts, or floods. But the African Department of the IMF in Washington is of the opinion that such food reserves distort the free market. In short: the state should play no role in the buying and selling of cereals, because that would violate the sacrosanct dogma of free trade. Since the great drought of 1985, which lasted five years, the rhythm of dry periods has accelerated. Famine now attacks Niger on average every two years.

Niger is a former French colony (1922–60), and today a French neocolony. It is the second-poorest country on the planet, according to the UNDP's Human Development Index. Vast resources lie untapped under its soil. After Canada, Niger is the second-biggest producer of uranium in the world. However, the AREVA group, a French public multinational industrial conglomerate (90 percent state-owned), has a monopoly on the development of the uranium mines at Arlit. The royalties paid by AREVA to the government in Niamey are ridiculously low.

In 2007, then president Mamadou Tandja decided to grant a permit to SOMINA (the Société des Mines d'Azelik) to develop the uranium mines at Azelik. The government of Niger would own 33 percent of SOMINA, while a Chinese company, Sino Uranium, would hold the remaining majority 67 percent of the company's shares. Tandja presented this as a done deal. At the same time, AREVA, which had been in Niger for forty years, was preparing to develop a site at Imouraren, south of Arlit. In early 2010, Tandja received a delegation from the Chinese minister of mines in the presidential palace. Niamey was rife with rumors that the Chinese too were interested in the mine at Imouraren.

Tandja's punishment was swift. On the morning of February 18, 2010, a military coup brought to power an obscure colonel named Salou Djibo, who broke off all discussions with the Chinese and reaffirmed Niger's "gratitude and loyalty" to AREVA. (In early 2011, free elections brought to power Mahamadou Issoufou, a former mining engineer and AREVA employee.)

Five years ago, the World Bank undertook a feasibility study on installing an irrigation system in Niger. The study concluded that installing groundwater pumps and a system of canals to channel river water would, without any major technical difficulties, permit the irrigation of 440,000 hectares (1.09 million acres). If it were realized, this project would therefore enable Niger to achieve self-sufficiency in food, and to permanently shield from famine 10 million people. Sadly, the world's second-largest uranium producer does not have a single penny to finance such a project.

The extreme poverty of the peoples who live in northern Niger, particularly the communities living at the foot of the Tibesti highlands, is the root cause of the Touareg rebellion, which has been rampant for the last ten years. Terrorist groups originating in Algeria, linked together in the network that calls itself Al-Qaeda in the Islamic Maghreb (AQIM), dominate the region. Their preferred action is taking Europeans hostage. They even kidnap Europeans in their favorite restaurant, Le Toulousain, in the center of Niamey, and in the heart of the whites' living quarters in the immense mining camp at Arlit. AQIM killers find willing recruits among young Touareg, who have been reduced by AREVA's policies to a life of permanent unemployment, despair, and poverty.

In southern Niger, in the Hausa territory around the city of Maradi in the ancient sultanate of Zinder, I witnessed the arrival of a devastating swarm of locusts. Far away, a strange sound fills the air, like a squadron of jets passing high overhead.

The sound comes closer.

Then, suddenly, the sky grows dark. Billions of migrating desert locusts—black and purple—beat their wings furiously. An enormous cloud obscures the sun. A sort of twilight dims your vision. The insects form a compact mass at the moment when they prepare to swoop down to earth. Their descent occurs in three stages. First they hover for several minutes, a restless, loud, threatening mass, above the villages, the fields, and the granaries they are preparing to attack. Then, with a terrifying roar, the mass drops halfway to earth. In countless numbers, they land on trees, shrubs, millet stalks, the roofs of huts, devouring everything their hungry jaws can seize. After a brief pause, the voracious army reaches the earth. Trees, bushes, millet plants, and other food crops are by now stripped bare, reduced to skeletons, the smallest leaf, the smallest fruit, the smallest grain having been devoured by the invader. The moving sea of locusts now covers the land for miles in every direction. On the surface of the soil, they devour every last edible morsel, churning the earth to a depth of half an inch.

The satisfied horde leaves as it arrived: suddenly, with a deafening noise, blocking out the sun. The farmers, their wives, their children emerge timidly from their huts, helpless to do anything but take stock of the disaster.

The size of the female locust ranges from about 7 to 9 centimeters (2¾–3½ inches), the males from 6 to 7.5 centimeters (2½–3 inches); they weigh 2 to 3 grams—about a tenth of an ounce. The locust devours in one day a volume of food equal to three times its own weight.

The desert locust is rife throughout the Sahel, the Middle East, North Africa, Pakistan, and India. Their ravenous swarms can cross oceans and continents. One neurotransmitter in particular, serotonin, unleashes their social instincts, leading to the formation of the swarm. Sometimes they join together to form superswarms comprising, it is said, several billion insects.

In theory, it is not difficult to fight these invaders: using all-terrain vehicles and small aircraft, powerful insecticides can be spread at the same time on the ground and in the air to deliver fatal doses of poison that penetrate entire swarms. By such means, during the locust invasion in 2004, Algeria rushed forty-eight vehicles into the field to spread 80,000 liters (21,000 gallons) of pesticides, Morocco used six vehicles to spread 50,000 liters (13,200 gallons), and Libya used six Toyota all-terrain vehicles to spread 110,000 liters (29,000 gallons). However, these pesticides are extremely toxic and can destroy the soil, making it unfit to plant for years. Handling such pesticides requires great care.

* * *

In the Bible, the following story is told in the book of Exodus. The Egyptian pharaoh refused to free the Hebrew people, whom he kept in slavery. To punish him, Yahweh sent a series of ten plagues against Egypt. First the waters of the Nile were changed into blood, killing all aquatic life. Then there followed plagues of frogs, lice or gnats, flies (or wild animals), pestilence that killed livestock, incurable boils, hail and thunder that destroyed orchards and crops, and locusts. After this came three days of darkness, and finally the deaths of Egypt's firstborn, both human and animal.

The eighth plague, the locusts, is described as follows:

> The locusts invaded the whole of Egypt and settled all over Egypt, in great swarms; never had there been so many locusts before, nor would there be again. They covered the surface of the ground till the land was devastated. They devoured whatever was growing in the fields and all the fruit on the trees that the hail had left. No green was left on tree or plant in the fields anywhere in Egypt.

Finally, Pharaoh gave in. He allowed the Jewish people to leave Egypt, and Yahweh put an end to the scourges with which he had ravaged Egypt.

In Africa, however, grasshoppers (locusts are the swarming phase of short-horned grasshoppers) continue to destroy crops, both in the field and after harvest. They return again and again, bringing famine and death.

The situation of Niger applies to all the countries in protracted crisis afflicted with this scourge. As a result, the levels of serious, permanent undernutrition are extremely high in all these countries, as the FAO table on pp. 36–37 clearly shows.

POSTSCRIPT 1: THE GAZA GHETTO

One of the most distressing of the world's current protracted crises does not show up in the FAO's table above. It is the direct consequence of the Israeli blockade of the Gaza Strip.

Figure 7. Levels of serious, permanent undernutrition

Country	Total population (in millions)	Number of undernourished people (in millions)	Percentage of undernourished people
	2005–7	2005–7	2005–7
Afghanistan	n/a	n/a	n/a
Angola	17.1	7.1	41
Burundi	7.6	4.7	62
Congo	3.5	0.5	15
Ivory Coast	19.7	2.8	14
Eritrea	4.6	3	64
Ethiopia	76.6	31.6	41
Guinea	9.4	1.6	17
Haiti	9.6	5.5	57
Iraq	n/a	n/a	n/a
Kenya	36.8	11.2	31
Liberia	3.5	1.2	33
Uganda	29.7	6.1	21
Central African Republic	4.2	1.7	40
Democratic Republic of the Congo	60.8	41.9	69
North Korea	23.6	7.8	33
Sierra Leone	5.3	1.8	35
Somalia	n/a	n/a	n/a
Sudan	39.6	8.8	22
Tajikistan	6.6	2	30
Chad	10.3	3.8	37
Zimbabwe	12.5	3.7	30

Percentage of children under age 5 who are underweight for their age	Mortality rate for children under 5 (percent)	Delayed growth (as percentage of body weight appropriate for age)
2002-7	2007	2000-2007
32.8	25.7	59.3
14.2	15.8	50.8
35	18	63.1
11.8	12.5	31.2
16.7	12.7	40.1
34.5	7	43.7
34.6	11.9	50.7
22.5	15	39.3
18.9	7.6	29.7
7.1	4.4	27.5
16.5	12.1	35.8
20.4	13.3	39.4
16.4	13	38.7
24	17.2	44.6
25.1	16.1	45.8
17.8	5.5	44.7
28.3	26.2	46.9
32.8	14.2	42.1
27	10.9	37.9
14.9	6.7	33.1
33.9	20.9	44.8
14	9	35.8

Gaza is a strip of land 41 kilometers (25 miles) long that reaches inland 6 to 12 kilometers (about 4 to 8 miles) from the eastern Mediterranean coast, bordered by Israel to the north and east and Egypt to the south. Gaza has been inhabited for about 3,500 years and is home to Gaza City, historically an important port and market town, a crossroads of exchange among Egypt and Syria, the Arabian Peninsula, and the Mediterranean.

Today more than 1.5 million Palestinians are crammed together in the 365 square kilometers (141 square miles) of the Gaza Strip, the great majority of them refugees and the descendants of refugees from the Arab-Israeli wars of 1947, 1967, and 1973.

In February 2005, the government of Prime Minister Ariel Sharon decided to evacuate its troops and officials from Gaza. Inside Gaza, the Palestinian Authority would from then on assume administrative responsibilities. But, in accordance with international law, Israel would remain in effect an occupying power: Gaza's airspace, territorial waters, and land borders would remain under Israeli control. On its side of the border with Gaza, Israel built an electrified barrier, fortified on both sides with mined areas, entirely surrounding the Strip. And Gaza became the biggest open-air prison on the planet.

As an occupying power, Israel has a duty to respect international humanitarian law and in particular to refrain from the use of hunger as a weapon against the civilian population. The facts on the ground are as follows.

I found myself one afternoon in Gaza City, in the sunshine-flooded office of the Commissioner-General of the United Nations Relief and Works Agency for Palestine Refugees in the Near East (UNRWA), Karen Koning AbuZayd. Born a Danish citizen and married to a Palestinian, she was wearing an elegant, loose Palestinian dress embroidered in red and black. Step by step, day after day, since the day in 2005 when she replaced her fellow Dane Peter Hansen, declared persona non grata by the occupying Israelis, she had fought against the Israeli generals to maintain the food distribution centers, hospitals, and 221 schools run by UNRWA.

The commissioner-general was preoccupied. "Anemia caused

by malnutrition—many of the children are sick with it," she told me. "We have had to close more than thirty of our schools. . . . Many children can no longer stand upright. Anemia is devastating them. They can no longer manage to concentrate. . . ." In a low voice, she continued, "It's hard to concentrate when the only thing you can think about is food."

After 2006, following the Israeli-Egyptian blockade of the Gaza Strip, the Gazans' food security situation deteriorated further. By 2010, the unemployment rate reached 81 percent of the able-bodied population. The loss of jobs, income, assets, and revenues has gravely endangered Gazans' access to food. Per capita income has been cut in half since 2006. In 2008, eight people in ten had an income below the official threshold of extreme poverty (less than $1.25 per day); 34 percent of the inhabitants were seriously undernourished. The situation is especially tragic for the most vulnerable groups, such as the 22,000 pregnant women, whose undernutrition will without question cause neurological problems among their unborn babies. In 2010, four in five Gazan families ate no more than one meal a day. In order to survive, 80 percent of the Strip's inhabitants depended upon international food aid. The entire population of Gaza is punished for acts for which they bear no responsibility.

On December 27, 2008, Israel's army, navy, and air force unleashed a broad-scale assault against the infrastructure and the inhabitants of the Gaza ghetto: 1,444 Palestinians, including 348 children, were killed, many with weapons that Israel was testing for the first time. The inhabitants of the ghetto were trapped, with no way to escape, caught between the electrified fence on the Israeli side and the locked-down border crossing at Rafah on the Egyptian side. More than 6,000 men, women, and children were injured, paralyzed, burned, or mutilated, or lost limbs.

The attacking Israeli forces systematically destroyed Gaza's civilian, and especially its agricultural, infrastructure. The Al Bader flour mill west of Jabalya, the biggest mill in Gaza and one of only three still functioning at the time, was attacked by Israeli F-16 fighter jets and totally destroyed. Yet bread is the staple food in Gaza.

Next, two successive attacks, on January 3 and 10, 2009, by jets armed with air-to-surface missiles, destroyed Gaza City's wastewater treatment plant, located on road number 10 in the al-Sheikh Ejlin neighborhood, as well as the dikes surrounding one of its wastewater lagoons. The city was thereby deprived of drinking water.

Richard Goldstone, president of the UN Fact Finding Mission on the Gaza Conflict, points out that neither the Al Bader flour mill, the water treatment plant, nor the farms at Ateya al-Samouni and Wa'ed al-Samouni in Zeytoun, where twenty-three members of one family were killed, were at any time ever used as shelter by Palestinian combatants. They could not therefore have constituted legitimate military targets.

As of 2011, the Gaza blockade continues. Despite the fall of the Mubarak regime in February 2011, Egypt continues to be an Israeli-American protectorate. In April 2011, the military council in power in Cairo opened the border at Rafah to allow people, but not merchandise, to cross the Egypt-Gaza border. The government in Tel Aviv allows just enough food to enter the ghetto to prevent a general famine, which would attract too much international attention. The Israelis are organizing undernutrition and malnutrition. Stéphane Hessel and Michel Warschawski believe that the goal of this strategy is to cause deliberate suffering among the inhabitants of Gaza so that they will rebel against the Hamas government. In order to achieve this political end, the Israeli government is using hunger as a weapon.

POSTSCRIPT 2: REFUGEES FROM THE NORTH KOREAN FAMINE

The UN Special Rapporteur on the Right to Food has, strictly speaking, no powers of enforcement. Nonetheless, I experienced some extraordinary moments, including one gray afternoon in November 2005 in New York. I was preparing to present my report before the Social, Humanitarian, and Cultural Affairs Committee (commonly known as the Third Committee) of the UN General Assembly. Waiting on the platform a few moments before my turn to speak, I felt a hand clutch at the sleeve of my jacket. A man was kneeling behind me, in such a way that he could not be seen from the assembly hall. He begged me, "Please, do not mention paragraph fifteen. . . . We have to talk."

It was the ambassador from the People's Republic of China. The paragraph of my report that he was so afraid of discussed the manhunts undertaken by the Chinese government against refugees from the North Korean famine. Two rivers—the Tumen, which forms part of the boundary between the two countries, and the Yalu, which runs along most of the border—are frozen for half the year, allowing thousands of refugees who dare defy the fiercely repressive North Korean government to cross into Manchuria, often only with great difficulty. This northeastern Chinese province has traditionally been home to a large community of the Korean diaspora. Men, women, and children are periodically arrested in Manchuria by the Chinese police and sent back to the authorities in Pyongyang. Many of the men repatriated by force are immediately shot; some disappear with the women and children into reeducation camps.

That same morning I had gone up to the thirty-eighth floor of the UN skyscraper, where the secretary general has his office. For five years, Kofi Annan had tried to negotiate the opening of camps on Chinese soil to receive North Korean refugees, which would be run by the UN. But Annan had utterly failed. That morning, the secretary general had given me the green light to condemn the Chinese manhunts.

Between 6 million and 24 million North Koreans are seriously undernourished. From 1996 to 2005, recurrent famine killed 2 million of the country's people. The Kim dynasty has built its nuclear arsenal over the mass graves of the victims of hunger.

As of early 2011, the situation in North Korea was once again catastrophic: floods destroyed rice fields, an epidemic of foot-and-mouth disease devastated livestock. Corruption, waste due to mismanagement, and the contempt for the starving that the terrorist Kim dynasty has shown complete the picture. Urgent action by the World Food Programme, with the support of certain NGOs (but not of the United States or South Korea), has attempted to contain the catastrophe. (The governments and NGOs that refuse to come to the aid of the starving people of North Korea justify their decision by explaining that they want to avoid the seizure of their aid by the authorities to feed the ruling class and the army.)

Amnesty International estimates the number of prisoners in North Korea at more than 200,000—including famine refugees sent back by the Chinese—confined in the country's political reeducation camps without trial or any prospect of being freed. Many returned famine refugees are held in camps such as Yodok in South Hamgyong province, one of six known political prison camps in North Korea. Yodok includes a "Total Control Zone" for those convicted of antiregime crimes, from which prisoners are never released. As Amnesty International reports:

> Family members of those suspected of crimes are also sent to Yodok. This can include parents, grandparents, sisters, brothers, nieces, nephews and cousins. Infants born in Yodok automatically become inmates, and if they are born in the "Total Control Zone," they will be there for life.

Thus entire families, covering several generations, including children of all ages, are incarcerated on the basis of "guilt by association."

Amnesty International has condemned the way that inmates designated as "troublemakers" in the camps, including children, are confined within cube-shaped concrete cells in which it is impossible to stand upright or to stretch out fully when lying down. Amnesty has noted in particular the case of a teenager who was held in such a cube for eight months. Furthermore, according to Amnesty, up to 40 percent of prisoners in the camps die of malnutrition. They attempt to survive the forced labor they are compelled to undertake (ten hours per day, seven days a week) by eating rats and grains foraged from animal feces.

The UN has shown itself to be powerless in the face of such horror.

4

THE CHILDREN OF CRATEÚS

Brazil's northeastern states account for 18 percent of the country's territory and are home to 30 percent of its total population. Most of the land belongs to the semiarid *sertão*, which stretches over 1 million square kilometers (386,000 square miles) of dusty, uncultivated savannah with scattered low thorn bushes, ponds dotted here and there, and a few rivers cutting across it. The sun burns white-hot, and the heat is scorching year-round.

Dressed in their leather outfits, the *vaqueiros* (cowboys) look after herds of several thousand head of cattle each, which belong to the *fazendeiros*, landowners with vast holdings who mainly descend from families that belong to Brazil's old Lusitanian vice-royalty, the colonial ruling class.

Crateús is a city in the *sertão* in the state of Ceará. The municipality covers an area of 2,000 square kilometers (770 square miles) and includes 72,000 inhabitants, most of them in the city itself. On the outskirts of the great *fazendas* (large estates, or *latifúndios*) and in the city's poverty-stricken suburbs stand the shacks of the *bóias-frias* (itinerant farm laborers) and their families: the landless rural labor force.

Every morning, including on Sundays, the *bóias-frias* gather in Crateús's central square. The *feitores*, the big landowners' overseers, walk among the crowd of starving men. They choose from among them the workers who will be hired for a day or a week, to dig an irrigation canal, fence a pasture, or do other work on a fazenda. Before a man leaves his hovel at dawn to sell himself in the square, his wife has prepared his lunch: a little rice, black beans, a potato. If he is lucky enough to be hired, her husband will eat his lunch cold at midday. If he is not hired, he will stay in the square, too ashamed to go home. Under a giant tree, he will wait, and wait, and wait some more. . . .

A *bóia-fria* in Ceará earns on average 2 reais a day, or about a dollar. In 2000, in the first term of President Luiz Inácio Lula da Silva, the government set the minimum wage for rural day laborers at twenty-two reais per day. But the *fazendeiros* of Ceará who respect laws made far away in the capital city, Brasília, are rare.

For several decades, Crateús has been the residence of an exceptional bishop: Dom Antônio Batista Fragoso. My very first visit to Crateús, in the early 1980s, when I was accompanied by my wife, Erica, had the air of a semiclandestine operation. Like Dom Hélder Câmara (1909–99), former archbishop of the cities of Olinda and Recife in the state of Pernambuco, Dom Fragoso was a determined partisan of liberation theology. In his sermons and in his social work, he defended the *bóias-frias*. The officers of the First Infantry Regiment of the Third Army stationed in Crateús and the great landowners in the area hated him. Many assassination attempts were organized against him. Twice the landowners' gunmen just barely missed their target.

Bernard Bavaud and Claude Pillonel, two Swiss priests affiliated with Dom Fragoso, had arranged our visit. And there we were at nightfall, at 1064 Rua Firmino Rosa, in front of the modest house that served Fragoso as the seat of his bishopric. Fragoso was a small, tough man from the northwest of the country, with a beaming smile. He welcomed us in perfect French. His warm simplicity immediately reminded me of the bishop in Victor Hugo's *Les Misérables*, the "Monseigneur Bienvenue" of the poor people of Digne.

The next day, Dom Fragoso drove us to a patch of wasteland a few miles beyond the last shacks at the edge of the city. "The anonymous children's burial ground," he told us. Looking more closely, we made out dozens of rows of little wooden crosses painted white. The bishop explained: according to Brazilian law, every birth has to be registered at the *prefeitura*, the city hall, but there is a fee to register a birth, and the *bóias-frias* did not have enough money to pay it. In any case, many of their children died shortly after birth as a consequence of fetal undernutrition or because their undernourished mothers could not breast-feed them. In short, said Dom Fragoso, "They come to earth to die."

Since they were not registered at the city hall, the children of the *bóias-frias* were unknown to the government, which could therefore not provide a permit for their burial. And without this government document, the Church could not bury the children in a consecrated cemetery. Dom Fragoso had found a solution on the edge of the law. With money from the bishopric, he had bought this bit of wasteland. Every week he buried there the "children who come to earth to die."

That morning, a friend of Bavaud and Pillonel's was with us: Cicero, a farmer who lived on a tiny patch in the middle of the *sertão*. He was a big man, with skin as dry as the surrounding countryside, like the skin of his wife and their many children, who all lived in a small house made of cut branches and adobe, where we would meet them all the next day. Then, he would show us the land he held as a *posseiro*, the holder of a legal title to a tiny property—barely 100 square meters (1,000 square feet)—where he was growing a few stalks of corn and where one pig rooted around. Cicero would tell us how, periodically, local landowners' *vaqueiros* sent their cows to graze on the land inside his fences, destroying his meager garden. He would tell us too that he was illiterate, which did not prevent him from listening to Radio Tirana, dreaming of revolution.

The sun was already high in the sky. Erica and I stood silent and still at the edge of the field dotted with small crosses. Cicero saw how moved we were. He tried to console us: "Here in Ceará, we bury these little ones with their eyes open so that they'll find the way to heaven more easily."

The sky in Ceará is beautiful, always strewn with a few pretty white clouds.

5

GOD IS NOT A FARMER

The macroeconomic situation, or in other words, the state of the world economy, is the ultimate determining factor in the struggle against hunger.

In 2009, the World Bank announced that, in the aftermath of the financial crisis of 2007–8, the number of people in "extreme poverty" (meaning, as I have said, those living on less than $1.25 per day) would rapidly increase by 89 million. As for "poor people," those living on less than $2 a day, their number would increase by 120 million. These predictions have been confirmed as millions of new victims of hunger have been added to the victims of normal "structural" hunger.

In 2009, the gross domestic product in every country in the world remained stagnant or fell for the first time since World War II. Worldwide industrial production tumbled by 20 percent. Those countries of the South that have most eagerly sought to integrate their economies within the world market are today the hardest hit: 2010 saw the biggest shrinkage of world trade in eighty years. In 2009, the flow of private capital to the countries of the South, and in particular to the "emerging" economies, fell by 82 percent. The World Bank estimates that, in 2009, the developing countries lost between $600 billion and $700 billion in capital investment. With global financial markets dried up, private capital is lacking.

In addition to this problem is the high level of debt owed by private companies, especially those in emerging countries, to Western banks. According to the UN Conference on Trade and Development (UNCTAD), nearly a trillion dollars' worth of debt came due in 2010. Given the insolvency of many of the companies based in countries in the South, this has caused a chain reaction leading to bankruptcies, factory closings, and waves of unemployment.

Another scourge has descended upon the poor countries: for many of them, transfers of foreign currency back home to their countries of origin by workers who have migrated to North America and Europe constitute an important part of their gross domestic product. In Haiti, for example, such transfers reached almost 49 percent of gross domestic product; in Guatemala, 39 percent; in El Salvador, 61 percent. Yet in North America and Europe, immigrants have been among the first to lose their jobs. Foreign-currency transfers to the developing world have thus drastically diminished or stopped altogether.

The speculative mania of the predators of the globalized financial industry cost, in total, $8.9 trillion in the industrialized Western nations in 2008–9. The Western nations have in particular spent trillions of dollars to bail out their delinquent bankers. But since the resources of their governments are not unlimited, their expenditures devoted to cooperative development ventures and humanitarian aid to the poorest countries have fallen dramatically. The Berne Declaration, a Swiss NGO, has calculated that the $8.9 trillion that the governments of the industrialized nations spent in 2008–9 to bail out their respective banks would equal seventy-five years of government development aid. The FAO estimates that for an investment of $44 billion in agricultural food production in the countries of the South over five years, the first of the UN's Millennium Development Goals could be realized: to halve, between 1990 and 2015, the proportion of people whose income is less than $1 a day; to achieve full and productive employment and decent work for all, including women and young people; and to halve, between 1990 and 2015, the proportion of people who suffer from hunger.

As I have said, only 3.8 percent of the arable land of sub-Saharan

Africa is irrigated. Like their forebears three thousand years ago, the vast majority of African farmers today practice a form of subsistence agriculture that relies on rain, with all the life-threatening unpredictability that implies. In a study released in May 2006, the World Meteorological Organization (WMO) examined the production of black beans in northeastern Brazil, comparing the productivity of an irrigated hectare and a nonirrigated one. The WMO's conclusion would hold equally true in Africa, and it is beyond dispute: rain-fed crops yield 50 kilograms per hectare, while irrigated crops yield 1,500 kilograms per hectare—a ratio of 1:30.

Africa, South Asia, Central America, and South America are all rich in strong ancestral farming cultures. Their farmers are culture bearers of impressive traditional knowledge, especially in meteorology. They have only to scan the sky to predict nourishing rains—or floods that will sweep away fragile sprouts. But, as I have said, their equipment is rudimentary: their principal tool remains the short-handled hoe. And the image of a woman or a teenager bent low over the short-handled hoe dominates the countryside from Malawi to Mali. There are no tractors. Despite the efforts of certain governments, such as Senegal's, to manufacture tractors domestically, or to import them en masse from Iran or India, there remain no more than 85,000 tractors in all of sub-Saharan Africa. As for draft animals, their number barely exceeds 250,000 head. The very low number of draft animals also explains the shockingly low usage of natural fertilizers.

Improved, high-yield seed, pesticides against locusts and grubs, mineral fertilizers, irrigation—all are lacking. The result is persistently very low productivity: 600 to 700 kilograms of millet per hectare in the Sahel in normal weather compared to 10 tons of wheat per hectare in European fields, as I have said. Moreover, to achieve even these low yields, the weather in the Sahel must be "normal"—that is, rain must fall as expected in June, moistening the soil and making it ready for sowing, and the "big rains" must in turn come in September—good, regular, constant rainfall that lasts for at least three weeks, soaking the young millet plants and enabling them to grow to ripeness.

Yet catastrophic weather patterns occur with increasing frequency. The "small rains" of June never come, the soil dries hard as concrete, and the seed remains on the surface of the cracked earth. The "big rains" come in great floods and, instead of gently watering the three-month-old plants, they "clean them out," as the Bambara say, uprooting them and sweeping them away.

The preservation of the harvest also poses an enormous problem. Each harvest should, in principle, enable a farmer and his family to survive until the next one. But today, according to the FAO, in the countries of the South, more than 25 percent of total harvests, including all products (grains, vegetables, and so on), are destroyed each year by the effects of climate, insects, or rats. Silos, as I have said, are rare in Africa.

Mamadou Cissokho cuts a figure that inspires respect. In his sixties, with a close-cropped cap of gray hair and strong features, a quick wit and ready, booming laughter, he is unquestionably one of the most widely listened-to leaders among the farmers of West Africa. A former teacher, he renounced his vocation at a young age and in 1974 returned to the village where he was born, Bamba Thialène, in Senegal, about 400 kilometers (250 miles) from the capital city, Dakar, to become a farmer. Since then, he has fed his large family as a subsistence farmer on a modest scale.

In the late 1970s, Cissokho joined together with all the farmers of the villages in the area around Bamba Thialène to found the country's first farmers' union. Since then, seed cooperatives have sprung up, first in the nearby region, then throughout Senegal, and finally in neighboring countries. The Réseau des Organisations Paysannes et des Producteurs d'Afrique de l'Ouest (ROPPA; Network of Farmers' and Agricultural Producers' Organizations of West Africa) was soon founded. ROPPA is today the most powerful farmers' organization not only in West Africa but on the entire continent. Cissokho is its director.

In 2008, the farmers' unions and cooperatives in countries in South, East, and Central Africa asked Cissokho to organize the Plateforme Panafricaine des Producteurs d'Afrique (Pan-African

Producers' Platform), a continent-wide union of farmers (raising both crops and livestock) and fishers that is today African producers' principal representative organization to the European Union in Brussels, African national governments, and the main intergovernmental organizations concerned with agriculture: the World Bank, the IMF, IFAD, and UNCTAD.

From time to time, I run into Cissokho in New York's John F. Kennedy Airport. He also often comes to Geneva. He works in Geneva with Jean Feyder, who since 2005 has been the courageous Permanent Representative of the Grand Duchy of Luxembourg to the United Nations, international organizations, and the World Trade Organization in Geneva. In 2007, Feyder was named president of the Committee on Trade and Development of the WTO, which attempts to defend the interests of the fifty poorest countries against the industrialized countries that control 81 percent of world trade. Since 2009, Feyder has also served as president of UNCTAD's executive council. In both these positions, he has made a certain modest farmer from Bamba Thialène his principal adviser. In the face of the most powerful forces in world agriculture, Cissokho assumed this role with determination, effectiveness—and humor. The battle against the inertia of African governments and intergovernmental organizations, and against the mercenary oligarchies of global finance, is a Sisyphean struggle. Between 1980 and 2004, the proportion of government development aid devoted to investment in agriculture, both multilateral and bilateral, fell from 18 to 4 percent.

But, to paraphrase British historian Eric Hobsbawm, nothing sharpens the mind like defeat. Every time I meet him, Cissokho's mind is even sharper than before. Fighting his way through interminable meetings in Geneva, Brussels, and New York with agrifood industry giants and the Western governments that serve their interests, Cissokho is nonetheless hardly an optimist. I have seen him recently exhausted, pensive, sad, worried. The title of the single book he has published sums up his current state of mind very well: *God Is Not a Farmer.*

6

"NO ONE GOES HUNGRY IN SWITZERLAND"

Jean-Charles Angrand, a historian from Réunion, once remarked to me, "The white man has taken the civilization of
the lie to levels never previously attained."

In 2009, the third World Food Summit brought together in the
FAO's headquarters in Rome on Viale delle Terme di Carcalla
a great many heads of state from the southern hemisphere, including Abdelaziz Bouteflika of Algeria, Oluṣẹgun Ọbasanjọ of
Nigeria, Thabo Mbeki of South Africa, and Luiz Inácio Lula da
Silva of Brazil. The Western heads of state who were their peers
made themselves conspicuous by their absence, with the exception of the prime minister of the host country, Silvio Berlusconi,
and the acting president of the European Union, who both put
in brief appearances. The total contempt thus expressed on the
part of the most powerful nations on the planet for a world conference that aimed to put an end to food insecurity, which afflicts
nearly a billion marginalized and undernourished people worldwide, shocked the media and public opinion in the countries of
the South.

Switzerland proclaims everywhere and to anyone who will
listen its commitment to the struggle against world hunger. Yet
the president of the Swiss Confederation, Pascal Couchepin, did
not deign to go to Rome. The government in Bern did not even

consider it worthwhile to send a federal councilor (as the cabinet ministers in the Swiss administration are called). Only the Swiss ambassador in Rome put in a quick appearance in the great meeting hall.

I have a friend in Bern who works in the agriculture department of the Swiss Federal Department of Economic Affairs. She is a former student of mine. She is a committed young woman, hardened by experience, who looks at the world with bitter irony. Disgusted, I called to talk to her. She said, "What are you so upset about? No one goes hungry in Switzerland." Nonetheless, one has to admit that the Western heads of state do not have a monopoly on indifference and cynicism.

In sub-Saharan Africa, 265,000 women and millions of infants die every year for lack of prenatal care. And when we study the global distribution of this phenomenon, we see that half of such deaths occur in Africa, while the continent's population represents only 12 percent of the world's population. In the European Union, governments spend on average 1,250 euros per person (about $1,650) annually on primary health care. In sub-Saharan Africa, the comparable expenditure ranges from 15 to 18 euros (about $20–24). At the most recent summit of heads of state of the African Union (AU), in July 2010 in Kampala, Uganda, Jean Ping, the president of Gabon and president of the executive committee of the African Union, set as the top item on the summit's agenda the fight against mother and infant undernutrition— much to his chagrin. François Soudan, editorial director of the magazine *Jeune Afrique*, followed the debates and reported them as follows:

> "Mothers and babies? But we're not UNICEF!" sputtered Muammar Gaddafi. The result: debate on this subject—a real chore—was dispatched in half an afternoon session by the heads of state, who were mostly daydreaming or half asleep. As for the journalists, pursued by the press attachés of NGOs trying desperately to sensitize them to their causes, they devoted to the debate only a small handful of reports

destined for the editorial wastebasket. The thing is, UA summits, you see, are only concerned with serious matters. . . .

The G8 and the G20 periodically hold summit meetings in
such places as Gleneagles, Scotland, or L'Aquila, Italy. At these
summits, the governments of the world's rich countries regularly
decry the "scandal" of hunger. Just as regularly, they promise to
release considerable sums to eradicate this scourge. At the instigation of British prime minister Tony Blair, the heads of state of the
G8+5, meeting in Gleneagles in July 2005, accordingly proposed
to spend immediately $50 billion to finance a plan of action to
fight extreme poverty in Africa. In his memoir, Blair discusses this
initiative at great length, and with evident pride. He sees it as one
of the three peak moments of his political career.

At the invitation of Silvio Berlusconi, the G8 heads of state met
in the small city of L'Aquila in central Italy in July 2009, which
had been struck three months previously by a terrible earthquake.
There, they unanimously approved a new plan of action against
hunger. This time they announced a commitment to spend
$20 billion immediately in order to support investment in subsis-
tence agriculture.

Kofi Annan was secretary general of the UN until 2006. The
son of Fante farmers in the high forest in the Ashanti region of
central Ghana, Annan has made the fight against hunger his lifework. An unassuming man who never raises his voice, sensitive,
often ironic, he spends most of his time today in Geneva. But he
regularly takes the shuttle between Founex, in the Swiss canton of
Vaud, and Accra, the capital of Ghana, where the headquarters
of the Alliance for a Green Revolution in Africa (AGRA), an organization of which he is president, is located.

Long familiar with the limitless hypocrisy of the Western powers, Annan agreed in 2007 to be president of a committee of
NGOs responsible for following up on the implementation of the
promises made at Gleneagles (officially the 2005 Gleneagles G8
Plan of Action). The result: as of December 2010, of the $50 bil-
lion promised, it appears that only $12 billion has in fact been

allocated and spent on financing various projects fighting hunger in Africa.

As for the promises of the G8 in L'Aquila, the situation is even darker: if the British weekly *The Economist* is correct, of the $20 billion in increased spending promised, only a fraction was in fact "new money," and little if any had been spent by late 2009. The magazine's editors sum up the situation soberly in the article's title: "If Words Were Food, Nobody Would Go Hungry."

7

THE TRAGEDY OF NOMA

In the previous chapters, I have discussed the effects of under-nutrition and malnutrition. But human lives can be ruined equally by the "diseases of hunger" that are among the consequences of these conditions. These diseases are legion, ranging from kwashiorkor (a syndrome caused by insufficient protein in the diet, with especially serious consequences in children, including permanent stunting of mental and physical development), to the blindness caused by lack of vitamin A (discussed in chapter 2), to noma, which ravages children's faces.

Noma's technical name is *cancrum oris* or gangrenous stomatitis. It is a gangrenous disease, a rapidly progressive, polymicrobial, opportunistic infection that develops in the mouth and leads to tissue destruction of the face. Its primary cause is malnutrition. Noma devours the faces of malnourished children, mainly those aged one to six.

Every living creature has in its mouth a great number of micro-organisms, constituting a heavy bacterial load. In well-nourished people with good basic oral hygiene, these bacteria are kept in check by the organism's immune system. But when prolonged under-nutrition or malnutrition weakens the immune system, these oral bacteria may become pathologically uncontrollable and break through the body's last immunological defenses.

The disease progresses in three stages. It begins with simple

gingivitis (inflammation of the gums) and the appearance in the
mouth of one or more ulcers (sores). If the disease is detected at
this stage, that is, within three weeks following the appearance of
the first oral ulcer, it can be easily cured: all that is necessary is to
cleanse the mouth regularly with a disinfectant, and to feed the
child appropriately, giving him access to the 800 to 1,600 calories
per day required, depending on his age, and the micronutrients
he needs (vitamins and minerals). His own immune system will
eliminate the gingivitis and the ulcers. If neither the gingivitis nor
the ulcers are detected in time, a wound forms in the mouth that
oozes blood. Gingivitis is succeeded by necrosis, or tissue death.
The child shakes with fever. But at this stage, all is not lost. The
treatment is still simple: it is enough to provide the child with antibiotics, adequate food, and rigorous oral hygiene.

One important expert on noma is Philippe Rathle, director of
the Winds of Hope Foundation, based in Lausanne, Switzerland;
Winds of Hope manages the No-Noma International Federation.
Rathle estimates that in total only 2 or 3 euros—$3 or $4—is
enough to provide ten days of treatment for noma at this stage,
when a child can still be cured.

If the child's mother does not have the small amount of money
necessary, or if she does not have access to medication, or if she is
unable to detect the wound in the child's mouth, or if she detects
it but, feeling ashamed, isolates the child, who cries and com-
plains ceaselessly, then the last threshold is crossed. Noma becomes invincible. First the child's face swells, then the necrosis
gradually destroys all the soft tissues. The lips, the cheeks disappear; a blackish furrow opens in the tissues, then deepens and
widens, leaving a yawning gap. The eyes drop down as the or-
bital bone is laid waste. Scar tissue deforms the face, shrinking
and sealing the jaws shut, making it impossible for the child to
open her mouth. The mother may now break the child's teeth on
one side so that she can pour, say, some millet soup in the child's
mouth in the vain hope that this grayish liquid will prevent her
child from dying of hunger. The child, with her ruined face and
frozen jaws, is unable to speak. With her disfigured mouth, she

can no longer articulate words, and can only emit groans and guttural noises.

Noma has four major consequences: disfigurement by destruction of the facial tissues, the inability to eat and speak, social stigmatization, and, in at least 80 percent of cases, death. The sight of a child's face devoured by noma, leaving the bones of the face visible, fills her family with shame and causes them to reject or hide her, which makes it even less likely that the medical care she needs will reach her. Death generally comes in the months that follow the collapse of the child's immune system, brought on by gangrene, septicemia, pneumonia, or bloody diarrhea. Fifty percent of afflicted children die within three to five weeks.

Noma can attack older children and, exceptionally, adults. The survivors live in agony. In most traditional societies of sub-Saharan Africa, the mountains of Southeast Asia, or the Andean highlands, the victims of noma are condemned as taboo, rejected as if their disease represented supernatural retribution, hidden from the neighbors. As BBC presenter Ben Fogle has said in a documentary on the disease, noma is like a punishment for a crime you haven't committed. Children who survive noma are withdrawn from society, isolated, walled off in their solitude, abandoned. They sleep with the animals.

The shame and the taboo of noma do not even spare the heads of state of countries afflicted with the disease. I confirmed this in May 2009 in the presidential palace in Dakar, Senegal, in the office of President Abdoulaye Wade. Wade is former dean of the law and economics faculty at the University of Dakar, a highly cultivated man, deeply informed about the problems his country faces. He was at the time the chairman of the Organisation of the Islamic Conference (OIC); together with the nations of the Non-Aligned Movement, the OIC, which includes fifty-seven states, forms a powerful bloc of votes in the UN. We talked about the OIC's strategies for working within the UN Human Rights Council (UNHRC). President Wade's analysis was, as always, brilliant and based on solid information. As I was about to leave, I asked him about noma, in order to elicit his thoughts on his responsibility for addressing the problem and to encourage him to put in place a national program to combat this terrible disease. Wade looked at me inquiringly: "But what are you talking about? I am not aware of this disease. There is no noma in our country."

In fact, I had just met that very morning, in the town of Kaolack, two representatives of Sentinelles, an NGO based in Switzerland that attempts to find children suffering from noma, to persuade their mothers to let them go to the local clinic or, in the most serious cases, to the university hospitals of Geneva or Lausanne. The Sentinelles workers gave me a precise picture of the extent of the disease, which is advancing not only along the Petite Côte, a section of Senegal's coast that includes both seaside resorts and fishing villages as well as the capital city, but in every rural area in Senegal.

Rathle, at Winds of Hope, estimates that in the Sahel, only about 20 percent of children afflicted with noma are detected. For survivors of noma, surgery is necessary. Volunteer surgeons at European hospitals in such cities as Paris, Berlin, Amsterdam, London, Geneva, and Lausanne, as well the rare doctor who travels to Africa to operate in the poorly equipped local clinics, accomplish miracles of reparative and reconstructive surgery, often of extreme complexity.

Two Dutch plastic surgeons, Klaas Marck and Kurt Bos, work in one of the only hospitals specializing in the treatment of noma in Africa, the Noma Children Hospital in Sokoto, Nigeria. They have learned many lessons from their experience: surgical procedures for car accident victims in particular have benefited from their work. Children suffering from noma have benefited, unquestionably. However, in order to reconstruct, even if only partially, the disfigured faces of these small children, as many as five or six successive operations are needed, all terribly painful. In many cases, only partial reconstruction of the face is possible.

As I write, I have before me on my desk photographs of little girls and boys, three, four, seven years old, with jaws sealed shut and drooping eyes. The images are horrible. Many of the children are trying to smile.

The disease has a long history, which Marck has pieced together. The symptoms were known in antiquity. The name *noma* was coined in 1680 by Cornelis van de Voorde, a surgeon in Middelburg, the Netherlands, who used it to describe a quickly spreading orofacial gangrene. In northern Europe, writings on the disease are relatively numerous throughout the entire eighteenth century. They associate noma with childhood, and with poverty and the malnutrition that accompanies it. Until the middle of the nineteenth century, noma was found all over Europe and North Africa. Its disappearance in these regions is due essentially to improvement in the social conditions of the populations involved, and to the reduction of extreme poverty and hunger. But noma reappeared massively in the Nazi camps between 1933 and 1945, especially in the concentration camp at Bergen-Belsen and the extermination camp at Auschwitz.

Every year, some 140,000 new victims are stricken with noma; about 100,000 are children of ages one to six living in sub-Saharan Africa. The proportion of survivors hovers around 10 percent, which means that more than 120,000 people die from noma every year. Children afflicted with noma are cursed: being born in general to gravely undernourished mothers, their own malnutrition begins in utero. Their growth is impaired even before they are born. Noma generally appears beginning with a mother's fourth child; she simply runs out of milk, weakened by her previous pregnancies. The larger her family, the more they have to share their food. The last children to be born suffer the most. In Mali, for example, slightly more than 25 percent of mothers manage to breast-feed their infants normally for the appropriate amount of time. The rest, the great majority, are too hungry to do so.

Another reason for the insufficient breast-feeding of hundreds of thousands of infants is premature weaning, the sudden early cessation of breast-feeding, which is mainly caused by pregnancies that fall too close together and by women's being compelled to undertake hard work in the fields. The cult of the large family is common all over the African continent. In rural areas especially, the status of women is tied to the number of children they bear. Separations, divorces, and the repudiation of wives are frequent and, accordingly, so is the separation of mothers from their small children. In many African societies, the father's family in fact keeps the children after the dissolution of a marriage, sometimes separating a mother from her infant even before the baby is weaned.

In their misfortune, Aboubacar, Baâratou, Saleye Ramatou, Soufiranou, and Mariam were lucky. These children from Niger, ages fourteen to sixteen, disfigured by noma, were living secluded in their homes in the heart of Zinder, the second-largest city in the country. Their families hid them, ashamed of the horrifying disfigurements that afflicted their offspring: a nose reduced to the nasal bone, cheeks with holes in them, ruined lips . . .

Sentinelles maintains a small but very active team in Zinder. After hearing about these children, two young women from Sentinelles visited their families. They explained that the children's disfigurement was not due to some kind of curse, but to a disease whose effects could be corrected, at least partially, with surgery. The families agreed to allow their children to go to Niamey, 950 miles away. A minibus took the children to the capital's national hospital, where Professor Jean Marie Servant and his team from the Saint-Louis hospital in Paris gave a human face back to each child.

Medical teams from France, Switzerland, the Netherlands, Germany, and other countries under the auspices of Médecins du Monde/Doctors of the World work in the hospital in Niamey three or four times a year for a week or two at a time. In other hospitals in Ethiopia, Benin, Burkina Faso, Senegal, and Nigeria, as well as in Laos, other doctors from Europe or the United States volunteer to operate on victims of noma.

Winds of Hope and the No-Noma International Federation accomplish tremendous results in detection, care, and reparative surgery for noma sufferers, and in the indispensable corollary of such work, fund-raising, as do other NGOs, such as SOS Enfant (founded by David Mort), Operation Smile, Facing Africa, Hilfsaktion Noma, and others. Yet while we must welcome the valuable contributions made by these NGOs and their doctors, their efforts nonetheless reach no more than a tiny minority of the children disfigured by noma.

Many NGOs are therefore trying to expand the effort to identify victims of noma and to finance reparative surgery in cases where it is possible. The Senegalese musician Youssou N'Dour, among other influential figures, has joined the fight by sponsoring such programs. However, it is clear that only WHO and the governments of countries affected by noma would be able to put an end once and for all to the suffering of children devastated by this horrible disease. Yet the indifference of WHO is as bottomless as that of the heads of state. In an incomprehensible decision, WHO has delegated the fight against noma to its African regional bureau. This decision is absurd for two reasons: first, noma is present not only in Africa but also in South and Southeast Asia and Latin America; and second, the African regional bureau of WHO has so far displayed an astounding passivity in the face of the suffering of hundreds of thousands of noma victims. The World Bank, whose charter includes a responsibility for fighting severe poverty and its consequences around the world, shows the same indifference. In an important 2003 paper, a team of German scientists showed that while "an excellent biological parameter for the presence of extreme poverty in a population is the structural presence of noma," there is an astonishing "lack of interest" in noma "from the side of public health policy makers" and global institutions such as "WHO and the World Bank." Many publications issued by WHO's Global Burden of Disease project do not even mention noma.

WHO initiates campaigns on its own authority against only two categories of diseases: contagious illnesses that pose a risk of becoming epidemics, and diseases for which a UN member state requests assistance. Noma is not contagious, and no member state has ever asked WHO for help in fighting it. In the capital city of each member state, WHO maintains a team comprising one WHO representative and a number of local employees. The team is permanently tasked with watching over health and sanitation in the country. WHO's representatives travel throughout the country, from urban neighborhoods to villages, hamlets, even nomadic encampments. They carry a checklist with details of all the diseases that they are expected to look out for. When a sick person is discovered, the local authorities are supposed to be alerted, and the patient taken to the nearest local clinic. But noma is not on WHO's checklist.

Together with Philippe Rathle and Ioana Cismas, my colleague on the Advisory Committee of the UNHRC, I appeared in Bern before the Swiss Federal Office of Public Health. The high-level official who met with us refused to present any resolution on noma to the World Health Assembly, with the following argument: "There are already far too many diseases on the checklist." WHO's representatives in the field are already overwhelmed. They hardly know what to do next. Adding another disease to the list—don't even think about it!

The coalition of NGOs led by Winds of Hope has drawn up an action plan against noma that focuses on preventive measures: training health workers and mothers to identify the first clinical signs of the disease, integrating noma into national and international epidemiological monitoring systems, and initiating ethological research into behaviors that lead to or prevent noma. Finally, it is necessary to ensure that antibiotics and supplies for emergency intravenous nutritional therapy are available in local clinics at the lowest possible prices. Realizing this action plan will cost money—money that the NGOs involved do not have.

Those who are fighting noma are caught in a vicious circle. On the one hand, there is the absence of noma from WHO's reports and checklists, and a lack of public attention to noma owing to the public's lack of scientific information on the extent and perniciousness of the disease. But on the other hand, so long as WHO and the ministers of health of the UN member states refuse to focus on this disease that affects the youngest and poorest people, there can be no extensive, in-depth research and no international mobilization against noma.

Moreover, noma is likewise of no interest to the pharmaceutical conglomerates, which exert a powerful influence over WHO, first because the medications used to treat the disease are inexpensive, and second because noma's victims have no money at all to pay for them.

Noma will only be finally eradicated in the world, as it was in Europe, when its causes, undernutrition and malnutrition, are rolled back for good.

PART II

THE AWAKENING OF CONSCIENCE

8

FAMINE AND FATALISM:
MALTHUS AND NATURAL SELECTION

Until the middle of the last century, hunger was a taboo subject, shrouded in silence. The graves of its victims sank into oblivion. Mass death by starvation was thought to be fated, inevitable. Like the plague in the Middle Ages, hunger was considered an invincible scourge that by its very nature lay beyond the power of human will to contain.

More than any other thinker, Thomas Malthus contributed to this fatalistic vision of human history. If the collective European conscience at the dawn of modernity remained deaf and blind to the scandal of millions of human beings dying from hunger, if Europeans even believed they could divine in this daily massacre a judicious form of demographic regulation, this is in large part owing to Malthus—and to his grand idea of "natural selection."

Malthus was born on February 13 or 14, 1766, and grew up at the Rookery, an estate near Westcott in Surrey, in southeastern England. His father lived off his inheritance as a "country gentleman."

On September 3, 1783, in a small townhouse on Rue Jacob in Paris, Ben Franklin, John Adams, and John Jay, as representatives of the American Congress of the Confederation, signed the Treaty of Paris with an envoy from King George III, acknowledging the independence of the United States of America. The loss of this North American colony had, for Great Britain, considerable repercussions. The landowning aristocracy, which had derived its income from American plantations and colonial trade, lost a great deal of its economic power and was supplanted by the rapidly growing industrial bourgeoisie. Immense factories were built, especially in the textile industry. From the marriage of coal and iron, the British steel industry was created. Millions of farmers and their families flocked to the cities.

Malthus took prizes as an undergraduate at Jesus College, Cambridge, graduating with honors, then completed his master's degree and later became a fellow of the college as well. In 1797, he took orders, and the following year became a country curate in the Anglican church at Okewood, near Albury in Surrey, not far from his childhood home.

However, Malthus discovered in London the appalling spectacle of extreme poverty. Uprooted rural workers who had joined the industrial underclass suffered from hunger. Having lost their bearings socially, many sank into alcoholism. Malthus would never forget the sight of mothers with pallid faces, haggard from undernourishment, of child beggars, and of prostitutes in wretched hovels. He was overcome by an obsession. How could these proletarian masses and their countless children be fed without endangering the food supply of society as a whole?

Even before the publication of his famous work, *An Essay on the Principle of Population*, the premises of Malthus's lifework are visible in an early unpublished pamphlet called *The Crisis* (1796), in which he observes that the principal challenge of his era is the problem of population and subsistence, since the needs of a growing population are constantly exceeding the available food supply. Already in this early work, Malthus asserts that it is an inevitable tendency of human beings, as of all living creatures, to increase in their numbers beyond the limits of their available food resources.

In 1798, Malthus published his famous *Essay*. Malthus revised the work through five more editions, enriching it with new material, confronting criticism, and even rewriting entire chapters to reflect his changing views. As late as 1830, four years before his death, when he reprinted a long extract from his article

"Population," which he had contributed to the supplement of the *Encyclopædia Britannica* in 1823, as *A Summary View of the Principle of Population*, Malthus was still revising.

The central thesis of the book revolves around a contradiction that Malthus considers insurmountable:

> Through the animal and vegetable kingdoms, nature has scattered the seeds of life abroad with the most profuse and liberal hand. She has been comparatively sparing in the room and the nourishment necessary to rear them. The germs of existence contained in this spot of earth, with ample food, and ample room to expand in, would fill millions of worlds in the course of a few thousand years. Necessity, that imperious all pervading law of nature, restrains them within the prescribed bounds. The race of plants and the race of animals shrink under this great restrictive law. And the race of man cannot, by any efforts of reason, escape from it. Among plants and animals its effects are waste of seed, sickness, and premature death. Among mankind, misery and vice.

For Malthus, a curate, "necessity, that imperious all pervading law of nature," was another name for God.

One of Malthus's most important statements of the "law of population" occurs not in the *Essay* but in his later *Principles of Political Economy* (originally published in 1820):

> Under this law of population, which, excessive as it may appear when stated in this way is, I firmly believe, best suited to the *nature* and situation of man, it is quite obvious that some limit to the production of food, or some other of the necessaries of life, must exist. Without a total change in the constitution of *human nature*, and the situation of man on earth, the whole of the necessaries of life could not be furnished in the same plenty as air, water, the elasticity of steam, and the pressure of the atmosphere. It is not easy to conceive a more disastrous present—one more likely

to plunge the *human* race in irrecoverable misery, than an
unlimited facility of producing food in a limited space. A
benevolent Creator then, knowing the wants and necessities of his creatures, under the laws to which he had subjected them, could not, in mercy, have furnished the whole
of the necessaries of life in the same plenty as air and water.
This shows at once the reason why the former are limited
in quantity, and the latter poured out in profusion. But if it
be granted, as it must be, that a limitation in the power of
producing food is obviously necessary to man confined to a
limited space, then the value of the actual quantity of land
which he has received, depends upon the small quantity of
labour necessary to work it, compared with the number of
persons which it will support; or, in other words, upon that
specific surplus . . . which by the laws of nature terminates
in rent.

Malthus's theory has prevailed ever since, and persists today in
public opinion. Some continue to believe that population inevitably grows unceasingly, while food and the land that produces it
are limited. Hunger reduces the number of people, guaranteeing
an equilibrium between fixed needs and available goods. From
an evil, God or Providence (meaning Nature) makes a good. For
Malthus the reduction of population by hunger is the only possible
way out that allows humanity to avoid a terminal economic catastrophe. Hunger thus amounts to a law of necessity.

Malthus's *Essay* includes, accordingly, virulent attacks against
the Poor Laws, the tentative efforts of the British government to
reduce by a rudimentary program of public assistance the terrible
suffering of the urban poor. Malthus asserts, in effect, that if a
man cannot live from his work, then that is simply too bad for him
and his family. He says that vicars should warn engaged couples
that if they marry and procreate, their children will have no help
from society. He claims that epidemics are necessary.

With each new edition of the *Essay*, the poor become more and
more Malthus's worst enemy. He says that laws designed to help
them are harmful; they only allow the poor to have more chil-

dren. Nature punishes man with want; the poor man must accept that the laws of Nature are the laws of God, and that they condemn him to suffer. And if church taxes crush the poor, too bad.

Such a theory must inevitably discriminate on racial grounds, and in the later editions of his book, Malthus indeed makes a survey of the world's peoples. Of the native peoples of North America, for example, he writes: "The tribes of hunters, like beasts of prey, whom they resemble in their mode of subsistence, will consequently be thinly scattered over the surface of the earth. Like beasts of prey, they must either drive away or fly from every rival, and be engaged in perpetual contests with each other."

An Essay on the Principle of Population met immediately with immense success among the ruling classes of the British Empire. Its arguments were debated in Parliament. The prime minister recommended reading it. Malthus's theses spread rapidly across Europe: Malthusian ideology admirably served the interests of the ruling classes and their exploitative practices. Malthus's ideas also made it possible to resolve another apparently insurmountable conflict: to reconcile the "nobility" of the bourgeoisie's civilizing mission with the famines and mass graves that it was causing. By adhering to Malthus's vision—accepting that while the suffering caused by hunger and the destruction of so many thousands of people were certainly dreadful, they were obviously necessary to humanity's survival—the bourgeoisie assuaged its own misgivings. The real threat was explosive population growth. Without the elimination of the weakest by hunger, the day would come when no human being on the entire planet would be able to eat, drink, or breathe.

Up until the mid-nineteenth century, Malthusian ideology ravaged the Western conscience, making Europeans for the most part deaf and blind to the suffering of the victims of hunger, especially in the colonies. The starving had become, in the ethnological sense of the term, taboo. Bravo, Malthus! Probably without clearly intending to, he had freed Westerners from their guilty conscience.

Except in cases of serious psychological derangement, no one can bear the sight of another human being dying from hunger. In naturalizing the mass death caused by starvation, by linking it to the idea of necessity, Malthus relieved Westerners of their moral responsibility.

9

JOSUÉ DE CASTRO, PHASE ONE

Suddenly, in the aftermath of World War II, the taboo was shattered, the silence broken—and Malthus relegated to the dustbin of history. The horrors of the war, Nazism, the extermination camps, the shared suffering and starvation of wartime: all led to an extraordinary reawakening of the European conscience. The collective conscience rebelled: *Never again!* This revulsion from war was expressed in a movement that sought a profound transformation of society, a movement through which people demanded independence, democracy, and social justice. The consequences were many and beneficial. Among other things, citizens required that governments create social safety nets for their people, but also that states create intergovernmental institutions, norms of international law, and weapons to fight the scourge of famine.

In his book *The Essence of Christianity*, the German philosopher Ludwig Feuerbach contends that only we human beings, alone among all animals, are capable of being aware of ourselves as members of our own species, and that this form of self-awareness is the very foundation of our moral sense. I would add that the consciousness of our shared human identity lies also at the foundation of the right to food. No one can tolerate the destruction from hunger of his fellow man or woman without endangering his own humanity, his very identity.

In 1946, the fifty-five member states of the UN, which had been founded the year before, meeting in Quebec City, created the Food and Agriculture Organization (FAO), the UN's very first specialized organization, with its headquarters in Rome. Its task: to develop subsistence agriculture and to oversee the equal distribution of food. On December 10, 1948, the member states of the UN, now numbering fifty-eight, during their general assembly in Paris, adopted the Universal Declaration of Human Rights, which affirms, in article 25, the right to food. Faced nonetheless with a series of the catastrophic famines, the UN's member states decided to go further in 1963 with the creation of the World Food Programme (WFP), tasked with providing emergency food aid.

Finally, in order to put more legal power behind the demand for the respect for human rights enshrined in the Universal Declaration, on December 16, 1966, the General Assembly adopted—separately, alas—two agreements, the International Covenant on Economic, Social and Cultural Rights (whose article 11 details the right to food) and the International Covenant on Civil and Political Rights. In the international context of the Cold War and given the resulting ideological divergences among the member states (capitalism versus communism), the second of these covenants was widely capitalized upon to denounce the violations against human rights in the countries of the Soviet bloc. As for the first covenant, the respect of states that are signatories to it for economic, social, and cultural rights has been overseen since the covenant's adoption by a committee of eighteen experts. Each signatory state must submit, upon ratifying the covenant and every five years thereafter, a report detailing the measures taken in the territory under its jurisdiction to satisfy the right to food.

At the end of the long night of Nazism, there was a dawning awareness of an obvious fact that would take years to spread among the world's peoples and their leaders: that the eradication of hunger is a human responsibility, and that hunger is no one's inevitable fate. The enemy can be beaten. It is enough to implement a certain number of concrete, collective measures to make the right to food real and enforceable.

It followed self-evidently, in the spirit of the instigators of the Covenant on Economic, Social and Cultural Rights, that the world's peoples could not leave the realization of the right to food up to the free play of market forces. Normative (that is, prescriptive, rights- and standards-based) interventions in markets would be indispensable, such as agrarian reform everywhere that unequal distribution of arable land prevailed; the public subsidy of staple foods for the benefit of those who cannot provide themselves with regular nourishment of adequate quality and in sufficient quantity; public investment in subsistence farming, at both the national and international levels, to provide fertilizer, irrigation, tools, and seed to ensure the preservation of arable soil and the enhancement of its productivity; equity in access to food for all; and the elimination of monopoly control over the markets in seed and fertilizers and over trade in staple foods by the multinational agri-food industry corporations.

One man contributed more than any other to the awakening of the conscience of the peoples of the West to the problem of hunger: the Brazilian doctor Josué Apolônio de Castro. I hope the reader will indulge me here in a personal recollection, of my meeting in February 2002 with Anna-Maria de Castro, his eldest daughter and intellectual heir.

Despite the deep eaves covering the little terrace of the Garota da Ipanema restaurant in Rio de Janeiro, the southern summer heat was suffocating. Looking down Rua Vinícius de Moraes, I could see the waves of the Atlantic glinting in the afternoon sun. The woman sitting across from me wore a serious expression. "The military believed they were finished with my father," she said. "But see now how he is coming back to us, and he is millions."

My meeting with Anna-Maria took place shortly after Luiz Inácio Lula da Silva took office as president of Brazil. Lula, a founding member of the Workers' Party (Partido dos Trabalhadores, known as the PT, a social-democratic party), was himself born into a very poor family in the interior of the state of Pernambuco, in the northeast, and lost two brothers to hunger as a child. One of his first decisions as president was to launch a national campaign called Fome Zero (Zero Hunger).

I was reminded of Josué de Castro's brilliant but tragic career. With his scientific work, his prophetic vision, and his militant actions, he profoundly marked his era. He broke the law out of necessity. He showed that hunger is a result of policies sustained by human beings, and that it can therefore be conquered, eliminated by human beings. The massacre is in no way inevitable. Combating hunger is a matter of bringing to light its causes and fighting against them.

De Castro was born on September 5, 1908, in Recife, the capital of Pernambuco, on the Atlantic coast, one of the country's largest cities. In Recife, a green ocean of sugarcane fields comes into view only a few miles from the city. The red earth of the Agreste—a narrow fertile zone that stretches the length of the coast, about sixty kilometers (forty miles) wide, through the states of Paraíba, Pernambuco, Alagoas, Sergipe, and Bahia, and beyond which lies the vast semiarid *sertão*—is a lost cause for such crops as beans, manioc, wheat, or rice. Like a ring of fire, fields of sugarcane encircle villages, market towns, cities. Sugarcane is the people's curse; its extensive cultivation precludes subsistence agriculture. As a result, to this day 85 percent of the food consumed in Pernambuco is imported, and infant mortality in the state is the highest in the region after Haiti. De Castro belonged in the deepest level of his being to that earth and to the people of the northeast. Like many of them, he was himself a *caboclo*, of mixed indigenous Brazilian and European ancestry.

By the time his book *Geografia da fome* (Geography of Hunger) was published in 1946, examining famine in Brazil and in the northeastern states in particular—that is, dealing with his own local and regional experience—de Castro already had a long career behind him. Armed with a degree in physiology from the medical school in Rio de Janeiro, he taught physiology, human geography, and anthropology at the University of Recife while practicing medicine at the same time. Like Salvador Allende, who was a pediatrician in Valparaiso, de Castro was exposed, in his private medical practice, at the hospital, and during house calls, to every facet of childhood undernutrition and malnutrition.

De Castro led an extensive, systematic series of precisely targeted inquiries, often under governmental authority, into the living conditions of thousands of families—*caboclos*, agricultural day laborers, cane cutters, tenant farmers, and *bóias-frias*—that enabled him to show that it was the system of large-scale plantation agriculture (the *latifúndios*) that was the cause of undernutrition and hunger. He also proved that it was not overpopulation of rural areas or cities that was responsible for the advance of hunger, but the opposite. Extremely poor people had large families out of anxiety about the future. Their children, of whom parents wanted as many as possible, constituted a form of life insurance. If the children survived, they would help their parents to live—and, above all, to grow old without dying of hunger. De Castro liked to quote a *nordestino* proverb: "The table of a poor man is meager, but the bed of poverty is fecund."

In *Documentario do Nordeste*, published in 1937, Castro writes, "If some of the *mestiços* turn out to be physically weak, afflicted with mental deficiencies and disability, this is not due to some social defect unique to their race, but to their empty stomachs. . . . Their misfortune comes not from their race, but from hunger. It is the absence of enough food that prevents their development and their functioning at full capacity. It is not the machine [i.e., the body] that is of poor quality. . . . Its labor yields little. With each step, it suffers. It breaks down early. . . . Owing to a lack of enough high-quality fuel."

Documentario do Nordeste pursued and amplified arguments de Castro had made in a short, slightly earlier text, *Alimentação e Raça*, which had appeared in 1935. Both books attacked the contention, then dominant among Brazil's political and intellectual elites, that Afro-Brazilians, the country's indigenous peoples, and the *caboclos* were all lazy, unintelligent, shiftless—and therefore undernourished—on account of their race. Brazil's white ruling classes were blinded by their racial prejudices.

The year 1937 witnessed the coup d'état of Gétulio Vargas and the establishment of his dictatorship and the regime of the Estado Novo (New State). The universalism of de Castro, then a young doctor, collided head-on with the fascist ideology and the proudly proclaimed racism of the ruling classes. In 1945, the defeat of the Axis powers inevitably led to the downfall of Vargas and the Estado Novo. Throughout this period, de Castro was invited by the governments of various countries to study the problems of food and nutrition; he traveled to Argentina (in 1942), the United States (1943), the Dominican Republic (1945), Mexico (1945), and, finally, France (1947).

De Castro's experience, at once both local and global, as we would say today, from the outset lent his scientific work, which includes some fifty titles, exceptional scope, complexity, and validity. Alain Bué, paying tribute to the man who was his supervisor and friend in France on the occasion of the hundredth anniversary of his birth, wrote, "The central argument of de Castro's entire oeuvre can be summed up in this observation: 'Whoever has money eats; whoever has none dies of hunger or becomes disabled.'"

Geografia da fome is the source of de Castro's most famous work, *Geopolítica da fome* (1951). In his author's preface to the revised and expanded 1965 edition of the French translation of *Geopolítica da fome*, de Castro explains that it was his editor at Little, Brown and Company in Boston who suggested that he extend to the entire world the methods that he had developed to support his analysis of Brazil, and which had given birth to *Geografia da fome*. *Geopolítica da fome* constitutes one of the major scientific works of the postwar period. The book met with universal acclaim all over the world. It was recommended by the newly created FAO, was translated into twenty-six languages, went through many reprints and editions, and had a profound impact on the awareness of hunger.

For the new book, De Castro's earlier title, *Geografia da fome*, which is rooted in the tradition of nineteenth-century descriptive social science, was changed to *Geopolítica da fome*; the author himself shows from the first chapter of the new book that hunger, if it must be in part related and attributed to geographical conditions, is in fact before anything else a political question. Hunger owes its existence and its persistence not to soil morphology but to human practices. It was in homage to Josué de Castro that I used his title as the subtitle of the original French edition of this book.

In his 1965 French preface, de Castro explains:

> But, although degraded by the Nazi dialectic, the word *geopolitics* retains its scientific value and should be rehabilitated with its real meaning. . . .
>
> The real meaning of the word *geopolitics* belongs to a scientific discipline that seeks to establish the correlations existing between geographical factors and phenomena of political character. . . .
>
> Few phenomena have so intensely influenced the political behavior of the [world's] peoples as the phenomenon of food and the tragic necessity of eating. Hence the harsh, living reality of a "geopolitics of hunger."

By treating the ravages of hunger as natural phenomena and invoking, to justify the mass death it causes, the "law of necessity," Malthus believed that he could shelter his conscience and that of the ruling classes from all concern for the hungry. De Castro made his readers aware, on the contrary, that persistent undernutrition and malnutrition profoundly disturb entire societies, both the starving and the well fed. As de Castro often said, "Half of Brazilians do not sleep because they are hungry. The other half do not sleep either, because they are afraid of those who are hungry." Hunger makes it impossible to build a peaceful society. In a state where a large part of the population is haunted by fear of tomorrow, only repression can ensure social peace. The institution of the plantation is the incarnation of violence. Hunger creates a permanent state of latent war.

De Castro frequently uses the term *artificial*. Undernutrition and malnutrition, he says, are "artificial" in the primary sense of the word *artifact*: they are phenomena created entirely by human activity, by, so to speak, experimental conditions. Colonization, monopolization of the soil, and monoculture are hunger's primary causes. They are responsible at the same time for low productivity and the unequal distribution of the harvest.

In many of his later works, de Castro would reinterpret the results of certain of his initial investigations in Pernambuco, as for example in his very fine *The Black Book of Hunger*. He would remain haunted throughout his life by the starving, toothless women, the children with their bellies swollen by worms, the cane cutters with empty eyes and broken wills of his natal Pernambuco.

Immediately following the end of the Estado Novo, with the reestablishment of a minimum of civil liberties, de Castro threw himself into political action against the *capitanias* (inland plantation estates) and the foreign multinational corporations that controlled the majority of agricultural production in Brazil. The country's agricultural production was at the time in large part intended for export—in this country of widespread hunger—and was enjoying dazzling growth compared to that of a drained and devastated Europe. After 1945, Brazil, where so many were suffering from hunger, became one of the world's biggest exporters of food. Together with Francisco Julião and Miguel Arraes de Alencar, de Castro organized *ligas camponesas*, peasants' or farmers' leagues, which were in effect the first farmers' unions in Brazil, to fight against the sugar barons, demand agrarian reform, and demand for cane cutters and their families the right to a regular supply of quantitatively and qualitatively adequate and sufficient food.

De Castro, Julião, and Arraes de Alencar were living dangerously. The landowners' gunmen, sometimes even the military police (who in Brazil exercise many functions of a civilian police force in, for example, the United States), attempted to ambush them on the chaotic roads of the São Francisco Valley and the ravines of the Capibaribe River. De Castro evaded numerous assassination attempts and continued with the fight. De Castro was the intellectual and theoretician of the group, Julião the organizer, and Arraes de Alencar the popular leader. The farmers' leagues were lent important support also from two priests of the *nordeste*, Dom

Hélder Câmara, who was in this period an auxiliary bishop in Rio de Janeiro and would later become archbishop of Recife and Olinda, and Father Italo Coelho, a native of Fortaleza, who will forever be remembered as the father of the poor in Copacabana.

In 1954, de Castro was elected the federal deputy for the PT (Workers' Party), Julião a deputy in the state of Pernambuco, and Arrães de Alencar governor of the state—*o goberandor da esperança,* "the governor of hope," as the people called him. In parallel with his work at the national level, de Castro played a decisive role on the international stage in the founding of the FAO. He was part of the small group of experts charged by the UN General Assembly with the preparatory work in advance of the FAO's founding, then Brazil's delegate to the FAO conference in Geneva in 1947, a member of the FAO's permanent consultative committee in the same year, and finally president of the FAO's executive council. First elected to the position in 1952, de Castro was reelected to this key post for a second consecutive term—an extraordinary exception to the organization's own rules—and served until 1955.

In these years of democratic hope and the quest for peace, de Castro was showered with prizes and honors. In 1954, the World Peace Council, then headquartered in Helsinki, awarded him its International Peace Prize. It was cold that day in Finland. De Castro lost his voice before the ceremony. Arraes de Alencar tells the story this way: Before the microphones and cameras, the theater filled with a colorful crowd of socialist notables and figures in the Finnish government, de Castro was struck by a coughing fit loud enough to shake the rafters of the hall. Finally he managed to utter a few words, just one sentence: "O primeiro direito do homem e de não passar fome" (The first human right is not to suffer from hunger). Then he sat down, exhausted.

De Castro was nominated three times for a Nobel Prize, once in medicine and twice for the Peace Prize. In the middle of the Cold War, he received the Franklin D. Roosevelt Prize of the American Academy of Political Science in Washington and, in Moscow, the World Peace Council's International Peace Prize. In 1957, he was awarded the Grande Médaille de la Ville de Paris, whose previous recipients include Pasteur and Einstein.

De Castro was perfectly well aware from experience of the often decisive influence exerted by agri-food industry conglomerates on national governments. He knew that no matter how many medals, prizes, and decorations governments showered upon him, they would never take any decisive action against hunger. He therefore pinned all his hope on civil society. In the Brazil of the *ligas camponesas*, he looked to the PT and the Movimento dos Trabalhadores Rurais Sem Terra as engines of change.

De Castro was also active at the international level. Starting in 1950, he traveled the world relentlessly: going to India, China, the countries of South America and the Caribbean, Africa, and Europe—wherever a governments, a university, a union asked him to speak. De Castro also co-founded the World Association for the Struggle Against Hunger (ASCOFAM), whose founding members included practically all the names of those who, after his death, would carry on the fight: Tibor Mende, René Dumont, Abbé Pierre, Father Louis-Joseph Lebret, Father Georges Pire (a future winner of the Nobel Peace Prize), and many others. In 1960, they succeeded in persuading the UN General Assembly to launch the first World Campaign Against Hunger. This campaign worked through schools, churches, parliaments, unions, and the media to inform and mobilize the public, and it met with a considerable response, principally in Europe.

Tibor Mende worked above all on the famines in China and India; his books include *China and Her Shadow*. Some of René Dumont's key early books, written during de Castro's lifetime, were directly inspired by him, including *Socialisms and Development*. As for Abbé Pierre, it was through the Emmaus movement, founded in 1949, that he spread de Castro's ideas. The Dominican priest Louis-Joseph Lebret deserves special mention. Among his colleagues in ASCOFAM, he was probably the closest to de Castro, and also his senior. It was Lebret who had de Castro's books first published in France. He was also the first to offer de Castro an academic position outside Brazil, at IRFED (Institut International de Recherche et Formation Éducation et Développement, the International Institute for Research and Training in Education and Development) in Paris, founded in 1958. Lebret's review *Développement et Civilisation* also often opened its pages to de Castro. Lebret was close to Pope Paul VI; one of the experts who contributed to Vatican II, he inspired the encyclical *Populorum progressio* (On the Development of Peoples), which played an important role in the struggle against hunger. A year before his death in 1965, Lebret was sent by the pope to Geneva to represent the Vatican at the first United Nations Conference on Trade and Development. Lebret mobilized progressive Catholics to join the struggle led by de Castro.

Today, more than 40 percent of the men, women, and children of Recife live in sordid shantytowns lining the Capibaribe. More than a million people live in homes without septic tanks or sewers, without running water or electricity, and in unsafe conditions. In their shacks made from sheet metal, wooden boards, or cardboard, famished rats bite and sometimes kill infants.

The Recife metropolitan area has one of the highest murder rates in Brazil, with 61.2 homicides per 100,000 inhabitants. The murder rate for children and adolescents is one of the highest in the world. Abandoned children number in the thousands. They are often the first victims of the death squads.

During my visits to Recife, I have gone out into the streets at night many times to accompany Demetrius Demetrio, the head of the Comunidade dos Pequenos Prophetas (the Community of Little Prophets), founded by Dom Helder Camara to rescue, feed, and provide day-to-day care for several dozen young street children, both boys and girls, from broken families. Some of the children I have seen were not even three years old; they were exposed to every danger, every kind of abuse and violence, every illness, and tormenting hunger. Those I met are surely all dead now, without having reached adulthood.

Deprived of work, the men and adolescents try to earn a few reais by hustling for biscate on the Avenida Conde de Boa Vista, which runs down to the harbor, lined with restaurants and tourist bars. *Biscate* is a term that covers all the odd jobs that make up the informal economy: sidewalk vending of ice cream, beverages, roasted peanuts, *cachaça* (sugarcane liquor), and *abacaxí* (pineapple); washing and guarding cars; shining shoes; and so on.

On the avenue, the *jangadas*, raftlike boats with triangular lateen sails, traditionally carved from whole logs and used in the open-ocean fishery, return to the docks at nightfall. The fish merchants wait in their vans. Haggard mothers and their starving children in rags wait in the shadows away from the streetlights. As soon as the vans have gone, these wretched people throw themselves eagerly upon the remains: fish heads or bones with a few scraps of flesh on them. The bones crunch in their mouths. I have witnessed this sight many times, heartsick.

Back when de Castro surveyed the shantytowns, about 200,000 people lived along the marshy banks of the Capibaribe. Over time, migrants from rural areas have spread out even over the water, building countless rudimentary homes on stilts. De Castro observed the astonishing way in which they fed themselves. The Capibaribe is a large river that descends from the hills of the coastal mountain range. Its waters are brown and turbulent in winter when, far away in the interior, the heavy rains and thunderstorms of July and August are unleashed. Most of the time, the river is a filthy cesspool where the inhabitants of the shantytowns relieve themselves, a vast stretch of marsh, almost unmoving, where crabs scuttle.

In his novel *Of Men and Crabs*, de Castro describes the "cycle of the crab." People relieve themselves underneath their shacks, in the river cesspool. The crabs, which are scavengers, feed on human waste along with other refuse thrown into the riverbed. Then the river dwellers, their legs stuck in the mud up to their knees, stir the muck to catch the crabs. They eat them, digest them, excrete them. The crabs feed on what the people excrete and throw away. The people catch the crabs and eat them. . . . And so the cycle continues.

10

HITLER'S "HUNGER PLAN"

Josué de Castro owed his victory over Malthus in part to Adolf
Hitler. Almost half of the 56 million combined civilian and
military deaths during World War II were caused by hunger and
its immediate consequences. For example, half the population of
Byelorussia died of hunger in 1942–43. Undernutrition, tuberculosis, and anemia killed millions of men, women, and children
across Europe.

Perhaps one of the most impressive chapters in *Geopolítica da fome*
is "Starving Europe," de Castro's account of hunger in Europe
before, during, and after World War II. In his section entitled
"Europe a Concentration Camp," De Castro writes:

> As soon as Germany invaded the various countries of
> Europe, she applied her policy of "organized hunger."
> According to Boris Shub, the master plan of the Third
> Reich was intended "to organize the pattern of privation for
> the peoples of Europe, apportioning among them in accor-
> dance with its military objectives, the short rations which
> remained after Reich priorities were satisfied."

A few pages later, de Castro summarizes the immediate results
of the implementation of the Nazi *Hungerplan*:

Such was Europe, then, stripped by the Nazi grasshoppers, devastated by bombs, paralyzed by panic, undermined by the fifth column as well as by administrative disorganization and corruption; and there starvation felt very much at home. Practically all the populations of Europe began a kind of concentration-camp existence. All Europe was one vast and somber camp.

The Nazi leaders, as is well known, were assisted by rigorous bureaucrats. Parallel to their system of racial discrimination, they created an equally punctilious system of discrimination in nutrition by dividing the populations of occupied countries into four categories:

- The "well fed," comprising population groups that fulfilled auxiliary functions in the Nazi war machine.
- The "underfed," comprising population groups that, on account of German requisitions of food, were limited to daily rations of 1,000 calories maximum per adult.
- The "hungry," a category encompassing population groups that the Nazis had decided to reduce in numbers, by restricting their access to food to a level below the threshold of survival. This category included most of the inhabitants of the Jewish ghettoes in Poland, Lithuania, Ukraine, and so on, as well as the Roma/ Sinti ("Gypsy") villages in Romania and the Balkans.
- The "starving," groups condemned to "extermination through hunger" or the "annihilation of superfluous eaters." In some of the Nazis' concentration camps and extermination camps, a starvation diet was used as a weapon of mass killing.

Hitler invested as much criminal energy in starving the peoples of Europe as he did in affirming German racial superiority. His hunger strategy had a double purpose: to ensure German self-sufficiency in food, and to subjugate the populations of occupied countries to the laws of the Reich.

Hitler was haunted by the food blockade that the British had imposed on Germany during World War I. As soon as he took power in 1933, he created the Reichsnährstand (RNS; Reich Food Corporation), a government body set up to regulate food production and distribution. By decree, all German farmers, food industries, stockbreeders, fisheries, and grain merchants were placed under RNS control. Hitler wanted war. He prepared for it by building up considerable stocks of food. A system of ration cards was imposed upon the people of Germany many years before the invasion of Poland. Between 1933 and 1939, the Third Reich absorbed 40 percent of all food exports from Yugoslavia, Greece, Bulgaria, Turkey, Romania, and Hungary. Before 1933, this figure had never exceeded 15 percent.

One of Hitler's first acts of outright theft took place at the Munich Conference in 1938. On September 29 and 30, Hitler met with Neville Chamberlain (representing the United Kingdom), Edouard Daladier (France), and Benito Mussolini (Italy), and blackmailed them into allowing the Reich to annex the Sudetenland, claiming that a majority ethnic German population there was being mistreated by the Czech government. Abandoned by the Western powers, Czechoslovakia was left to Hitler's mercy. The Führer then forced the Prague government to sell to Germany, according to the terms of a duly signed commercial contract, 750,000 tons of cereals—which were never paid for.

One war was declared, Hitler organized the systematic plunder of food in the occupied countries. The conquered countries were ransacked, their food reserves stolen, their agriculture, livestock, and fisheries placed in the exclusive service of the Reich. The experience accumulated over the course of the previous seven years by the RNS proved invaluable. With thousands of railway cars at its disposal, as well as thousands of agronomists, the RNS systematically bled dry the food production sectors of France, Poland, Czechoslovakia, Norway, Holland, Lithuania, and so on. Robert Ley was head of the Third Reich's German Labor Front (in effect, the minister of labor). The RNS fell under his jurisdiction. Ley declared in a statement in 1940, "'A lower race needs less room, less clothing, less food' than the German race." The Nazis called the pillage of the occupied countries "war requisitions."

Poland was invaded in September 1939. Hitler immediately annexed the cereal-producing plains of western Poland and put them under the direct administration of the RNS. The region was separated from the Nazi-installed General Government, which controlled the rest of occupied Poland, and incorporated into the Reich as a Reichsgau (administrative subdivision) called the Wartheland (after the Warthe River—or the Warta in Polish—which crosses it; the area was also called the Warthegau). Early in the winter of 1939, the farmers and stockbreeders of the Wartheland were forced to deliver to their new masters, without any compensation, 480,000 tons of wheat, 50,000 tons of barley, 160,000 tons of rye, more than 100,000 tons of oats, and tens of thousands of head of livestock (cattle, hogs, sheep, goats, and chickens).

But the looting was equally effective in the General Government area of Poland. It was the governor-general of the territory himself, Hans Michael Frank (magisterially described by Curzio Malaparte in his book *Kaputt*), who organized the requisitions program. In the course of 1940 alone, Frank extracted from occupied Poland and shipped to the Reich 100,000 tons of wheat, 100 million eggs, 10 million kilograms (more than 22 million pounds) of butter, and 100,000 hogs. Famine set in throughout the Wartheland and the rest of Poland.

Of all the European countries occupied by the Nazis, two took exceptional precautions against hunger early in the war: Norway and the Netherlands. Norway had known a horrifying famine during the Napoleonic Wars, caused by Napoleon's Continental Blockade against the British Isles. By the outbreak of World War II, the country had built the world's third-largest merchant marine. The government in Oslo bought food all over the world. Along the fjords in the far north, the Norwegians stockpiled in warehouses tens of thousands of tons of dried and salted fish, rice, wheat, coffee, tea, and sugar, and thousands of hectoliters of cooking oil. The Dutch did the same. When the Nazis invaded Poland,

the government in the Hague proceeded to make emergency purchases of food all over the world, putting in reserve 33 million
chickens and adding 1.8 million hogs to the country's livestock.

When the Nazi armies surged across Dutch and Norwegian
territory, the RNS officials who followed in their wake could not
believe their eyes: they had based their planned requisitions on
out-of-date data. Now, to their delight, they discovered unex-
pected treasures. They stole everything.

The Nazis invaded Norway in 1940. Three years later, the
Norwegian economist Else Margrete Roed made a preliminary
assessment of the Norwegian food situation, which de Castro
quotes at length:

> One day the Germans suddenly appeared, brutally invaded
> the country and took possession of all [the Norwegians'] re-
> serves. "They descended upon us like locusts and devoured
> everything in their way. Not only did we have three or four
> hundred thousand greedy Germans to feed in Norway; the
> German transports which brought them to us sailed back
> laden with Norwegian food and other foods." From then
> on . . . all these products, one after another, gradually dis-
> appeared from the market: "First eggs, then meat, wheat
> flour, coffee, cream, milk, chocolate, tea, canned fish, fruits
> and vegetables, and finally cheese and fresh fish—all disap-
> peared down the German mouths."

In the Netherlands as in Norway, tens of thousands of people
died of hunger or its consequences. Kwashiorkor, anemia, tuberculosis, and noma ravaged the children of both countries.

Practically all the occupied countries endured the same suffering. In many countries, deficiencies in animal protein rose
precipitously. The estimated amount of protein required per
adult in occupied Europe varied—according to the country, the
Nazis' population categories, and the arbitrary whim of the local
Gauleiter (leader of a regional branch of the Nazi party)—from 10
to 15 grams (about one-third to one-half ounce) per day. Likewise,

the consumption of fats also plummeted: in Belgium, for example, from 30 grams per adult per day to only 2.5 grams.

In the racial hierarchy established in Berlin, the Slavs occupied one of the bottom rungs of the ladder, just above the Jews, the Roma/Sinti, and blacks. The rationing of food was thus even crueler in Eastern Europe. The daily ration for an adult civilian in the occupied countries in the east fell rapidly below a thousand calories (2,200 calories per day is an average adult's daily requirement). Consisting mainly of rotten potatoes and bread that was often stale and moldy, the civilian ration was soon equal in calories to that of the concentration camp prisoners. Maria Babicka succeeded in smuggling an exposé out of Poland in 1943, which was published in the United States; "the Polish people were reduced, Babicka says, 'to eating dogs, cats, and rats, and to making broth from the skin of dead animals and the bark of trees.'" In the winter of 1942, Babicka notes, the average daily ration of a Polish adult fell below 800 calories. Hunger edemas, tuberculosis, an almost complete inability to work normally, and progressive weakness and lethargy due to anemia tormented the Poles.

The Nazi strategy for weakening or destroying entire peoples or certain groups by hunger comprised several variants. Heinrich Himmler's Reichssichertshauptamt (Central Security Department), for example, had conceived a scientific plan for the annihilation by hunger of selected groups of people judged to be "life unworthy of life" (*Lebensunwertes Leben*): the *Hungerplan*. The executioners of the Reichssichertshauptamt were determined to hunt the Jews and the Roma/Sinti to death. They used every weapon—the gas chambers, mass executions—and also the weapon of hunger. Thus, everywhere in occupied Europe, from the Baltic to the Black Sea, hermetically sealed Jewish ghettos, surrounded by walls and "protected" by cordons of SS men and enclosing in some cases hundreds of thousands of people, were subjected to a regime of starvation; many of the ghettos' inhabitants died of hunger.

I am reminded of my visit to the former concentration camp of

Buchenwald, in Thuringia, before Germany's reunification. The prisoners' barracks, the quarantine hospital, the execution chamber (execution was by pistol shot to the neck of a prisoner seated and handcuffed to a chair), the SS quarters, the two crematorium ovens, the muster ground where selected prisoners were hanged daily, the brick villa built for the commandant and his family, the chimneys, the kitchens, the mass graves—all are situated on an idyllic hillside that you ascend on foot, through a forest of beech trees, after leaving the small city of Weimar, below in the valley, where, until his death in 1832, the poet Goethe once lived and worked. Immediately upon entering the camp, after passing through the gray iron gates, now rusted, you find yourself in a vast enclosure, as big as a soccer field, surrounded by barbed-wire fences ten feet high. My guide, a young East German, explained to us in a neutral voice, "This is where the authorities"—he says "the authorities" and not "the Nazis"—"starved prisoners to death. . . . The camp was first used in 1940 with the arrival of Polish officers."

Several hundred Polish officers were imprisoned at Buchenwald. They had to take turns sleeping, because the camp could barely contain them all standing upright. The prisoners spent their days and nights on their feet, jammed together. They were deprived of any food and given only a little brackish water that fell drop by drop from two iron pipes. They had no protection from the elements: no shelter, no blankets. They were brought to Buchenwald in November, with only their uniform greatcoats to protect them. The snow fell on their heads. It took them two or three weeks to die. Then a new batch of Polish officers arrived. The SS had set up machine-gun posts all around the barbed wire. Escape was impossible.

Historian Timothy Snyder has taken advantage of the opening of the archives of the former Eastern Bloc countries since the Soviet Union's disintegration in 1991. He describes the suffering endured by Soviet prisoners of war condemned by the Nazis to "extermination through hunger." The Nazi butchers were ferocious accountants. Every camp, whether it relied on forced labor, extermination by gas, or forced starvation, had to maintain its *Lagerbuch* (accounts book). In many of these *Lagerbücher*, the SS relate with relish and in vivid detail recurring cases of cannibalism. They see in the cannibalism to which young Soviet prisoners of war resorted as they were dying of hunger the ultimate and definitive proof of the Slavs' barbaric nature. The archives also reveal that in one of the camps that practiced extermination through hunger, many thousands of Ukrainian, Russian, Lithuanian, and Polish prisoners of war signed a petition, which they presented to the SS commandant. They demanded to be shot.

The blindness of the Allied High Command throughout the war to the Nazis' strategy of control and destruction through hunger of selected groups within the occupied countries astounds me. At Buchenwald, I was struck by the single railway line, its tracks covered in grass and wildflowers, which in almost bucolic fashion winds its way through the charming, gently rolling Thuringian countryside. Not one American, English, or French bomber ever tried to destroy it. The trains full of deportees continued to arrive day by day at the foot of the hill, as if it were the most normal thing in the world. Some of my friends have visited Auschwitz, and they returned with the same feeling of revulsion and incomprehension: again, a single railway line bringing its daily quota of prisoners to this factory of death remained perfectly intact until early 1945.

In the autumn of 1944, the Allied forces liberated the southern part of the Netherlands. They then pressed their advantage eastward to penetrate German territory, leaving the entire northern part of Holland, and in particular the cities of Rotterdam, the Hague, and Amsterdam, under the iron rule of the Gestapo. Members of the Resistance were arrested by the thousands. Hunger ravaged entire families. The national railway system was paralyzed. Winter came. Hardly any food could be brought into the cities from the countryside.

Journalist Max Nord, in his introduction to the 1947 book of photographs *Amsterdam tijdens de Hongerwinter* (Amsterdam in the Hunger Winter), wrote: "The western part of Holland lived in bitter despair, in the greatest penury, without food or fuel." So many people died that, since "timber for coffins was scarce, there were long rows of deceased to be seen in the churches." He adds, bleakly, "the Allied forces marched toward Germany, paying us no heed."

Throughout World War II, Stalin, like Hitler, distinguished himself by committing massacres though starvation. Adam Hochschild points by way of example to one freezing night in February 1940 when the NKVD (the Soviet secret police) arrested 139,794 Poles, the parents and wives and children of Polish prisoners of war, who were then deported in cattle cars to Siberia. The NKVD was able to detain entire families because the occupying Soviet troops in eastern Poland had allowed captive Polish officers and men to correspond with their families; in this way, the secret police were able to learn the soldiers' families' home addresses. Since the camps of the Gulag were already overpopulated, the police had to decide to "liberate" thousands of families, who were abandoned along the way, without food, without blankets, without water. As Hochschild recounts, all along the railway lines in the Soviet Far East, as far as the Pacific, groups of these people were "scattered" and left to die of hunger.

11

A LIGHT IN THE DARKNESS: THE UNITED NATIONS

In Europe, the ordeal of hunger did not end with the surrender of the Third Reich on May 8, 1945. European agriculture was ravaged, its economies in ruins, its infrastructure destroyed. In many countries, people continued to suffer from hunger, malnutrition, and illnesses caused by the lack of food and aggravated by the immune system collapse of entire populations.

Josué de Castro observes in this regard:

> One of the toughest postwar problems was to provide food for a Europe that had been torn and broken by six years of fighting. Several factors had led to a marked decline in food production, and now stood in the way of recovery. Particularly responsible for the decrease of food production in Europe were a decline in productivity of the soil, due to lack of manure and fertilizers; a reduction in cultivated area; a relative scarcity of agricultural labor; and a shortage of farm tools and machinery. Acting together, in most cases, these factors had reduced agricultural production 40 per cent below the prewar level. This decrease had still graver effects on Europe's food balance because the population of the continent, in spite of heavy loss of life, had increased by some 20 per cent during the war.

"The case of France," writes de Castro, "is typical":

> In this country, the war, the occupation and the libera-
> tion led to extremely unfavorable conditions for food sup-
> ply. France continued to go hungry for a long time after
> the liberation, and was shamefully preyed upon too by
> the corruption of the black market. The agricultural re-
> covery of France encountered serious obstacles, outstand-
> ing among them being the extremely poor condition of her
> farm lands and the absolute lack of mechanized agricul-
> tural machinery.

One of the most difficult problems to resolve affecting food production was the lack of fertilizers. In France, the amount of mineral fertilizers available had reached 4 million tons in 1939; by
1945, this figure had fallen to a quarter million tons.

Another problem was the agricultural labor force. More than
100,000 French farmers had abandoned their fields between 1939
and 1945, either because their farms had been devastated or because the occupying Nazis had ruined them financially. Moreover,
during the war, 400,000 French farmers had been taken prisoner,
and 50,000 killed. Recovery was slow and painful. As De Castro
writes:

> As a result of the tremendous drop in production, and the
> absolute lack of financial resources to buy foods outside the
> country, France was forced to go through long years of nu-
> tritional poverty after the war. It was only with the aid of
> the Marshall plan that the country succeeded in emerging
> from this economic asphyxia, and that the people were able
> little by little, to return to a more tolerable diet.

The suffering, deprivation, undernutrition, and hunger endured
by Europeans throughout the dark years of Nazi occupation made
them receptive to de Castro's ideas. Rejecting the Malthusian
ideology of the law of necessity, they committed themselves

wholeheartedly to the campaign against hunger and to building international organizations responsible for leading the fight.

The personal destiny of Josué de Castro and his battle against hunger are intimately linked to the growth of the United Nations.

Today, the UN is a bureaucratic dinosaur directed by the passive and colorless Ban Ki-moon of South Korea, incapable of responding to the needs, expectations, or hopes of the world's peoples. The UN arouses hardly any popular enthusiasm. This was not, however, the case when the UN was created in the aftermath of the war. The name of the United Nations emerged for the first time in 1941, when the very words *united nations* aroused strong emotions. And the idea was explicitly linked to the fight against hunger.

On August 14, 1941, the British prime minister, Winston Churchill, and the American president, Franklin D. Roosevelt, met at Roosevelt's instigation on the American cruiser USS *Augusta* off the coast of Newfoundland. Their meeting, known since as the Newfoundland Conference, yielded a document called the Atlantic Charter, whose principles Roosevelt had anticipated several months before. In his State of the Union address delivered on January 6, 1941, referred to ever since as the Four Freedoms speech, Roosevelt had proposed four fundamental freedoms that people "everywhere in the world" ought to enjoy, and which he said he was dedicated to achieving: freedom of speech and expression, freedom of worship, freedom from want, and freedom from fear. The idea of the Four Freedoms had already been at the heart of the New Deal, the program of reform that had carried Roosevelt into the White House in 1932.

The four freedoms lie at the foundation of the Atlantic Charter, whose fourth and sixth articles read in full:

> 4. They [the United States and Great Britain] will endeavor, with due respect for their existing obligations, to further the enjoyment by all States, great or small, victor or vanquished, of access, on equal terms, to the trade and

to the raw materials of the world which are needed for their economic prosperity.

6. After the final destruction of the Nazi tyranny, they hope to see established a peace which will afford to all nations the means of dwelling in safety within their own boundaries, and which will afford assurance that all the men in all the lands may live out their lives in freedom from fear and want.

In 1941, hunger was still tormenting entire populations in the countries occupied by the Nazis and consumed by the war. Once military victory had been won, it was clear to Churchill and Roosevelt that the UN must as one of its first priorities mobilize all its resources and all its energies in support of the eradication of hunger.

Sir John Boyd-Orr (who was at the time consultant director to the British Imperial Bureau of Animal Nutrition) wrote the following year:

When the fighting forces of the Axis Powers have been completely defeated, the United Nations will be in control of the whole world. It will be a shattered world. In some countries the political, economic, and social structures will be almost completely destroyed. Even in countries least affected by the war, they will be badly damaged. It is obvious that the world will need to be rebuilt. . . .

The opportunity is there, but the immensity of the opportunity is equaled by the immensity of the task. The task cannot be accomplished unless the free nations which have united in the face of the common danger of Nazi world domination remain united to co-operate in building "the new and better world."

A few months before his death, Roosevelt reaffirmed the principles announced on board the USS *Augusta*:

We have come to a clear realization of the fact that true individual freedom cannot exist without economic security and independence. "Necessitous men are not free men." People who are hungry and out of a job are the stuff of which dictatorships are made.

In our day these economic truths have become accepted as self-evident. We have accepted, so to speak, a second Bill of Rights under which a new basis of security and prosperity can be established for all regardless of station, race, or creed.

The global campaign against hunger, inspired in large part by the scientific work and the tireless activism of Josué de Castro and his comrades, was buoyed up by the energy and hope that Roosevelt so magnificently expressed.

Two inherent limits to this great project must be mentioned here. The first concerns the world's political organization in this period. The member states of the UN—whose composition was very much at issue in the early 1940s, before the UN's founding in 1945—were dominated by the United States, Western Europe, and the Commonwealth: that is, by Western and mainly white countries. At the end of World War II, two-thirds of the planet's people still lived under the yoke of colonialism. Only fifty nations participated in the United Nations Conference on International Organization in San Francisco, where the United Nations Charter was drawn up and signed on June 26, 1945. In order to be admitted to the founding conference, a country's government had to have declared war against the Axis before May 8, 1945. When the UN convened the General Assembly in Paris on December 10, 1948, to adopt the Universal Declaration of Human Rights, there were fifty-eight member states, as I have said above.

The second limit to the UN's effectiveness is rooted in a contradiction within the organization that dates back to its founding: its legitimacy lies in the free adherence of the world's nations to the UN Charter, as is set out in the charter's preamble. But the UN itself is an organization of *states* (comprising representatives of their *governments*), not of *nations*. Every component of the UN is governed by a different array of its member states. Its executive is the Security Council, which today numbers fifteen states. The General Assembly, which constitutes the UN's parliament, today comprises 193 states. The UN's Economic and Social Council (ECOSOC) oversees the UN's specialized organizations, such as the FAO, WHO, the WMO, the International Labour Organization (ILO), and so on; ECOSOC comprises ambassadors, that is, representatives of its own fifty-four member governments. The UN Human Rights Council, which is responsible for monitoring the implementation of the Universal Charter by the UN's member states, comprises representatives of forty-seven states. As we all know, moral convictions, enthusiasm, and the spirit of justice and solidarity do not inhere in the idea of the *state*. The state's *raison d'être* is *raison d'état*. These intrinsic limits to the UN's effectiveness persist today.

Nonetheless, the aftermath of the war did indeed witness a great awakening of conscience and consciousness in the West, which broke the taboo on speaking of hunger. The peoples who had endured famine would no longer accept the doxa of fatalism and inevitability. Hunger, they well knew, was a weapon that the occupying powers had used to break and destroy them. They knew it from experience. They would now resolutely commit themselves to the fight against this scourge, behind de Castro and his comrades.

12

JOSUÉ DE CASTRO, PHASE TWO:
A VERY HEAVY COFFIN

In Brazil in 1961, João Goulart, the candidate representing the Workers' Party (PT), was elected president of the republic. He immediately initiated a series of reforms, including, as a first priority, agrarian reform. He appointed Josué de Castro as Brazil's ambassador to the UN agencies in Geneva. It was there that I knew de Castro. At first glance, he was the very picture of a bourgeois gentlemen from Pernambuco, down to the discreet elegance of his clothes. Behind the fine lenses of his glasses his eyes shone with an ironic smile. He voice was soft. He was warm but discreet, very friendly, obviously a deeply moral man. De Castro had proven to be effective and conscientious as the head of the Brazilian mission, but little inclined to diplomatic socializing. His two daughters, Anna-Maria and Sonia, and his son, also named Josué, attended a public school in Geneva.

De Castro's appointment to the mission in Geneva unquestionably saved his life. After President João Goulart was overthrown in a coup d'état masterminded by the Pentagon on the night of March 31–April 1, 1964, and Field Marshal Humberto de Alencar Castelo Branco took power as president on April 15, Brazilian democracy was destroyed and a period of military dictatorship inaugurated that would last twenty-one years. Indeed, at the top of the first list of enemies of the state issued by the coup leaders were the names of leading figures of the PT, including João Goulart, Leonel Brizola, Francisco Julião, Miguel Arrães de Alencar, and Josué de Castro. At dawn on April 10, paratroopers took over the governmental palace in Recife. Arrães de Alencar was already at work. He was abducted and disappeared, but soon a great wave of international solidarity compelled his captors to release him; like de Castro, he had become, throughout Latin America, a symbol of the fight against hunger.

Arrães de Alencar was forced into ten years of exile, first in France, then in Algeria. I saw him, not for the first time, in 1987. With the end of the dictatorship in 1985, he had been reelected governor of Pernambuco. He had immediately picked up work where he had left off twenty years before. In a husky voice, barely audible, he told me, "I have found all the old problems—multiplied by ten."

As for Julião, he went underground on the morning of the coup. Denounced, he was arrested in Petrolina, on the border of the states of Pernambuco and Bahia. Despite appalling torture, he survived and was freed. He died in exile in Mexico. (Goulart and Brizola successfully evaded arrest, escaping to Uruguay.)

From 1964 to 1985, Brazil's military dictatorship—barbaric, cynical, and efficient—ravaged the country, as a succession of generals and marshals, each bloodier and stupider than the ones before, governed a marvelous and stubbornly resistant people. In Rio de Janeiro, air force secret-service torturers held sway downtown, in hangers at Santos-Dumont Airport. Others, affiliated with the marines, tormented kidnapped students, professors, and union leaders in the basement of the marines' general-staff headquarters, a vast white building eight stories high located a few hundred meters from the Praça Quinze, the square at the heart of historic downtown Rio, and Cândido Mendes University. Every night, army commandos, disguised as civilians and armed with lists of suspects, roamed throughout the city, from the middle-class neighborhoods of Flamengo and Botafogo, to the chic Copacabana, to the endlessly sprawling, wretched suburbs of the Zona Norte, where working-class neighborhoods stretch out into a sea of shacks on stilts in the favelas.

But all across Brazil, from the Amazon delta to the Uruguayan border, there was active resistance to the dictatorship. Although the farmers' leagues, agricultural and industrial unions, political parties, and leftist movements were all annihilated by the secret services and the dictatorship's commandos, a few armed groups survived to wage a clandestine war against the generals in the countryside, such as the VAR-Palmares, of which the current president of Brazil, Dilma Rousseff, was a member.

Fourteen countries offered to welcome de Castro. He chose France. In Paris de Castro was one of the founders of the Centre Universitaire Expérimental in Vincennes (today the Université de Paris VIII in Saint-Denis), where he taught starting in the beginning of the academic year in 1969.

Although he was living in exile, de Castro did not disappear from the international stage. Despite the opposition of the generals in power in Brasilia, the UN continued to offer him a platform to speak out. In 1972, de Castro delivered an address at the first United Nations Conference on the Human Environment (also known as the Stockholm Conference). His arguments in favor of family subsistence agriculture, as the only form of agriculture exclusively in the service of human needs, provided a powerful source of inspiration to the final resolution and action plan of this, the UN's first ever conference on the environment.

De Castro died of a heart attack in his Paris apartment on the morning of September 24, 1973, at the age of sixty-five. His funeral service took place in the Église de La Madeleine. His children having negotiated—with difficulty—their father's return to Brazilian soil, the airplane carrying his body landed in Guararapes Airport in Recife, where an immense crowd awaited its arrival. But no one was able to reach the coffin. The surrounding area was ringed with thousands of riot police, paratroopers, and soldiers. Such was the influence of the deceased, and his place in Brazilians' hearts: the dictators feared his coffin like the plague. De Castro today lies interred in the São João Batista Cemetery in Rio de Janeiro.

***

André Breton once wrote:

> Everything leads to the belief that there exists a certain
> point in the mind beyond which life and death, the real
> and the imaginary, past and future, the communicable and
> the incommunicable, high and low cease to be perceived as
> contradictory.

The life of Josué de Castro confirms this hypothesis. Born
Catholic, he was not a practicing Christian. But he remained a
believer—a believer beyond dogmas.

De Castro and Gilberto Freyre shared a stormy relationship,
but one marked by mutual respect. Freyre, the scion of a wealthy
family in Recife who lived in a house called the Casa Amarela,
was the author of the famous book *The Masters and the Slaves*, and
a conservative who supported the military dictatorship—at least
until Institutional Act #5 (AI-5), promulgated at Christmas in
1968, abolished the last vestiges of democratic freedom.

Freyre was the protector of the best-known Umbanda house in
Recife, the Terreiro de Seu Antônio, in the neighborhood known
as Coque. (Umbanda is a syncretistic Afro-Brazilian religion,
similar to Candomblé, which blends together various African and
indigenous Brazilian traditions with Catholicism and Kardecism,
the beliefs of the nineteenth-century French systematizer of
Spiritism, Allan Kardec.) A passionate sociologist, de Castro entirely shared the point of view of Roger Bastide, who believed that
the task of the sociologist is to "explore all the ways that humans
have of being human." In de Castro's day, all the Afro-Brazilian
religions, which had persisted since slavery, including Umbanda
and Candomblé, were regarded with great contempt, imbued
with racism, by the white ruling classes. De Castro was fiercely
interested in the cosmogonies and other beliefs of such popular
cults. Encouraged by Freyre, he faithfully attended ceremonies at
the Umbanda house that Freyre supported.

I encountered the Terreiro de Seu Antônio in the early 1970s
thanks to Roger Bastide. The tropical night was thick with all the

odors of the earth. The faraway sound of drums rolled like muf-
fled thunder across the sky. We had to walk for a long time down
alleyways without streetlights, full of restless shadows, through
the vast Coque quarter. The guard recognized Bastide. He called
Seu Antônio. Bastide persuaded him, at great length, to allow me
to enter. Before the altar young black men and women, all dressed
in white, spun interminably, obsessively, round and round, until
they fell into a trance and the house resounded, over the heads of
the silent witnesses, with the voice of Xango.

The universe of Umbanda is full of mysteries, strange chance
events, logical coincidences. Should one see signs in what follows?

On January 17 and 18, 2009, the Université de Paris VIII
celebrated its fortieth anniversary. Paris VIII is surely, after the
Sorbonne, the best-known French university abroad, and the
one with the most prestigious reputation in the countries of the
South. In the words of its president, Pascal Binscak, it is "a world
university." Born out of the student revolt of May 1968 and in-
carnating the student movement's spirit of openness and radical
critique, Paris VIII has conferred since its founding more than
two thousand PhDs, about half of them to men and women from
Latin America, Africa, and Asia. Garcia Alvaro Linera, the current vice president of Bolivia; Marco Aurélio Garcia, a special foreign policy adviser to the Brazilian president; Fernando Henrique
Cardoso, the former president of Brazil; and Cardoso's wife, Ruth
Cardoso, all either taught or studied at Paris VIII.

The school decided to celebrate its anniversary with an international colloquium dedicated to Josué de Castro on the hundredth anniversary of his birth. I was invited to speak at the event
and received that day—at the instigation of Alain Bué and his
colleague Françoise Plet—an honorary doctorate.

And then this. It was Olivier Bétourné, at the time a young editor at Éditions du Seuil, who ensured the republication in France
of de Castro's *La géographie de la faim* in the early 1980s. And it was
also precisely Bétourné, today the president of Éditions du Seuil,
who had the idea for this book you are now reading—to reignite
the struggle.

PART III

ENEMIES OF THE RIGHT TO FOOD

13

THE CRUSADERS OF NEOLIBERALISM

For the United States and its hired guns—the World Trade Organization, the International Monetary Fund, and the World Bank—the right to food is an aberration. For them, the only human rights are civil and political rights.

Standing behind the WTO, the IMF, and the World Bank, the U.S. government and its traditional allies, to be sure, are the gigantic global private corporations. The growing control of these corporations over vast sectors of food production and trade clearly has considerable repercussions for the exercise of the right to food.

Today, the top two hundred corporations in the agri-food industry control about a quarter of the world's food production resources. Almost all of these companies realize astronomical profits and have at their disposal financial resources much greater than those of the governments of the countries in which they do business. They exercise monopoly control over every aspect of the entire food supply chain from production to distribution, including the sale, processing, and marketing of food products, the effect of which is to restrict the choices of both farmers and consumers.

Since the publication of Dan Morgan's classic book *Merchants of Grain*, the American media has used Morgan's term "merchants of grain" to designate the principal global agri-food corporations. Yet Morgan's term is inadequate: the giants of agri-food control not only the setting of prices and the trade in foods but also all the sectors essential to the food industry, especially the production of seed, fertilizers, and pesticides, as well as storage, transportation, and so on.

Just ten corporations, including Aventis, Monsanto, Pioneer, and Syngenta, control one-third of the market in seed, with estimated annual sales in 2010 of $23 billion, and 80 percent of the market in pesticides, with estimated annual sales of $28 billion in the same year. Ten other corporations, with Cargill prominent among them, control 57 percent of the sales of the world's top thirty retailers and represent 37 percent of the revenues earned by the top one hundred manufacturers of foodstuffs and beverages. Six companies control 77 percent of the market in fertilizers: Bayer, Syngenta, BASF, Cargill, DuPont, and Monsanto.

In certain sectors of agricultural-products processing and marketing, more than 80 percent of the trade in a single product lies in the hands of a few oligopolies. As Denis Horman has shown, "six companies control some 85 percent of world trade in cereals; eight divide up about 60 percent of world sales in coffee; three control more than 80 percent of sales in cocoa; and three share 80 percent of the trade in bananas." The same lords of the food-industry oligarchies dominate the bulk of transportation, insurance, and distribution of foodstuffs. On the commodities exchanges trading in agricultural products, their traders fix the prices of the most important staple foods. As Doan Bui notes, "From seeds to fertilizers, from storage to processing to final distribution . . . they lay down the law for our planet's millions of farmers, whether they are farmers in Beauce or smallholders in Punjab. These businesses control the world's food."

In his pioneering book published fifty years ago, *Modern Commodity Futures Trading*, Gerald Gold used the terms *cartel* or *monopoly* to designate such companies, depending upon the sector of economic activity he examined. Today, the UN uses the term *oligopoly* to better characterize markets in which a very small number (*oligo-* in Greek) of producers or sellers confront a very large number of buyers. As João Pedro Stedilé writes of the agri-business giants, "Their purpose is not to produce food, but to produce merchandise and make money."

* * *

Let us examine more closely the paradigmatic example of Cargill. Cargill is active in sixty-six countries, with 1,100 branches and 131,000 employees. In 2007, the company had $88 billion in sales and net profits of $2.4 billion, a 55 percent increase in net profit over the previous year. In 2008, a year of global food crisis, Cargill achieved sales of $120 billion and a profit of $3.6 billion.

Cargill is one of the companies most closely watched by NGOs, especially American organizations. Consider, for example, the report issued by Food and Water Watch entitled *Cargill: A Threat to Food and Farming*. As the report explains, Cargill, especially through its subsidiary Mosaic, was until recently, among other things, the world's largest producer of mineral fertilizers. Because of its quasi-monopoly, Cargill was able to make prices rise considerably in 2009; the prices of nitroglycerine-based fertilizers rose 34 percent and the prices of phosphate- and potash-based fertilizers doubled.

In 2007, the year for which the most recent figures are available, Cargill was the second-biggest meat packer, the second-biggest feedlot owner, the second-biggest pork packer, the third-biggest turkey producer, and the second-biggest producer of animal feed in the world. Regarding its treatment of meat products, Food and Water Watch writes:

> Cargill has been a major advocate for technological fixes to food safety challenges that could also be addressed through more stringent sanitation and other preventative measures. Only days before the November 2007 recall of hamburger patties, a Cargill representative testified before Congress and claimed its use of carbon monoxide in meat packaging helped inhibit the growth of *E. coli*. There is no evidence that carbon monoxide hinders or inhibits the bacteria that cause foodborne illness, and the FDA did not approve it for that use. The company had treated much of the beef involved in the recalls with carbon monoxide, which is primarily used in meat packaging to keep meat looking fresh and red long after it may have spoiled. . . .

Food & Water Watch views the use of carbon monoxide in food packaging as consumer deception. It makes it impossible for customers to use visual cues alone to determine if meat is fresh.

Cargill also uses the highly controversial technology of food irradiation to kill bacteria. As Food and Water Watch writes, irradiation

creates chemical byproducts in the food, some of which are known carcinogens and some of which are unique to irradiated food and have been linked to tumor promotion and genetic damage. In scientific studies irradiated food has been shown to cause premature death, stillbirths, mutations, immune system failure, and stunted growth in animals.

Because of its harbor facilities and silos around the world, Cargill is able to stockpile enormous quantities of corn, wheat, soybeans, and rice—and to wait for prices to rise. Conversely, because of its fleet of ships and cargo aircraft, Cargill can liquidate its merchandise in record time. As Food and Water Watch writes:

By 2008, millions of people around the globe faced starvation that spawned rioting and instability due to skyrocketing food prices, and Cargill was making billions of dollars in profit. . . .

For agricultural communities in the developing world, high prices for imported food like the corn and wheat that Cargill sells coincided with low prices for the tropical crops like cotton and cocoa that Cargill buys from these communities. . . . Even during the recent commodity price surge, tropical cash crop prices grew modestly while food staple prices doubled or tripled. Between January 2006 and June 2008, the world price for coffee, tea, cotton and bananas grew by a third or less, while rice prices tripled, corn and soybean prices grew by more than 150 percent and wheat prices doubled. This price-spread benefits Cargill, but puts food beyond the reach of many rural communities in the developing world. The United Nations estimated that an additional 130 million people worldwide became malnourished because of the high price of food during the 2008 food crisis.

Cargill is also one of the most powerful cotton merchants in the world. Its principal sources of cotton are in East Asia, and particularly Uzbekistan. Cargill United Kingdom maintains a purchasing office in Tashkent that buys about $50 million to $60 million worth of cotton per year. In its Country Reports on Human Rights Practices, the State Department in Washington has for several years condemned the use of compulsory child labor in Uzbekistan's cotton fields. By some estimates, as many as a quarter million children are forced by the government to work for paltry wages in the annual cotton harvest; children who fail to meet their daily quotas are often severely beaten.

Cargill and the other oligopolies play, at certain moments, a key role in the explosion of food prices. When the market is rising, cargoes of food commodities may change hands multiple times even while they are being shipped. Analyzing the speculative bubble in grain prices of 1974, Morgan writes that

> speculative fever gripped the grain trade as prices began to climb. Cargoes were changing hands twenty of thirty times before they actually were ready for delivery. Cargill might sell to Tradax, which might sell to a German merchant, who would sell to an Italian speculator, who could hand it off to another Italian, who would pass it on to Continental.

Today, Cargill, through a financial services and commodity-trading subsidiary, is active in the world's principal agricultural commodities exchanges. As Food and Water Watch writes:

> Cargill also participated in the commodity speculation that helped propel the [2008] food crisis, both through its dominant market position in the cereal market and the activities of its commodity futures trading subsidiary. Cargill operates a financial services and commodity-trading

subsidiary that trades financial instruments (like interest rate and currency swaps) and energy futures as well as
farm commodities. This allows Cargill to manage its own
purchases and sales of farm products but also to act as a
financial services firm for other investors to speculate on
commodity prices. In 2008, excess speculation on the commodity markets helped to drive up food prices and significantly contributed to the food crisis.

One of the greatest sources of power exercised over markets by
such conglomerates, which are like octopuses of world trade with
tentacles extending in all directions, is vertical integration. Jim Prokopanko, acting as a spokesman for the conglomerate, describes what
he calls total control of the food production chain, using as an example the "chicken supply chain." Cargill produces phosphate-based
fertilizer in Tampa, Florida. With this fertilizer, Cargill grows soy
in fields in the United States and Argentina. In Cargill factories, the
soybeans are transformed into flour. In ships that belong to Cargill,
this soy flour is then sent to Thailand, where it is used to fatten
chickens on large farms that also belong to Cargill. The chickens
are then killed and gutted, in an almost entirely automated process, in factories that belong to Cargill. Cargill packs the chickens.
The Cargill fleet then carries them to Japan, the Americas, and
Europe. Cargill trucks distribute the chickens to supermarkets,
many of which belong to the MacMillan and/or Cargill families,
which control 85 percent of the multinational conglomerate's stock.

On the international market, the oligopolies use all their power
to set food prices to their advantage—that is to say, as high as possible. But when it comes to conquering a local market by eliminat-
ing local competitors, the merchants of grain willingly practice
outright dumping. One example: the total ruin of local poultry
production in Cameroon. Massive imports of cheap foreign poultry products threw tens of thousands of poultry-farming and egg-
producing families into deep poverty.

As soon as the local producers had been destroyed, the lords of
agribusiness raised their prices.

* * *

Private multinational corporations in the food industry often exert a decisive influence on the policies of international organizations, as they do on the policies of virtually all the Western governments. And these corporations act as determined adversaries to the right to food.

They argue as follows: Hunger does in fact constitute a scandalous tragedy. It is due to the insufficient productivity of global agriculture, since the amount of goods available does not meet existing needs. To fight against hunger, we must therefore increase productivity, an objective that can be attained only if two conditions are met: first, the most intensive possible industrialization of food production processes, mobilizing the greatest possible capital investment and the most advanced technology (such as transgenic seed, high-performance pesticides, and so on), and, as a corollary, the elimination of the myriad family and subsistence farms deemed "unproductive"; and second, the most complete possible liberalization of the world market in agricultural goods. Only a totally free market is capable of extracting the maximum from the economic forces of production. That is the credo. Any normative intervention in the free play of the market, whether it be undertaken by states or by intergovernmental organizations, can only inhibit the development of the forces of production.

The policies of the United States and of the intergovernmental organizations that uphold those policies constitute a threat pure and simple to the right to food. I must, however, admit that these policies proceed from neither blindness nor cynicism. In the United States, officials are perfectly well informed about the ravages of hunger in the South. Like all other civilized countries, the United States claims to fight hunger. But, according to the U.S. government, only the free market will be able to overcome this scourge. Once global agricultural productivity reaches its maximum potential through liberalization and privatization, universal access to quantitatively and qualitatively adequate and sufficient food will follow automatically. Like a shower of gold, the market, liberated at long last, will pour out its blessings on humanity.

But the market can also malfunction, the Americans admit.

Catastrophes can always happen—a war, a climatic disturbance. In such a case, international emergency food aid will rescue the afflicted.

Today it is the WTO, the IMF, and the World Bank that determine the economic relations that the dominant countries maintain with the peoples of the South. But in matters of agricultural policy, these organizations faithfully obey the diktat of private multinational corporations. This is why the FAO and the WFP, which were originally founded to combat extreme poverty and hunger, no longer play, in comparison to the corporations, anything but a vestigial role.

In order to gauge the abyss that separates the enemies and the supporters of the right to food, consider the positions taken by various countries regarding the UN's Covenant on Economic, Social and Cultural Rights and the obligations that follow from it. The United States has always refused to ratify the covenant. The WTO and the IMF have fought and continue to fight it. The states that are signatories to the covenant undertake three distinct obligations. First, they must "respect" the right to food of the inhabitants of their own territories, which means they must not do anything that would interfere with this right.

Take the example of India. The country's economy is heavily dependent on agriculture to this day: 70 percent of the population lives in rural areas. According to the United Nations Development Programme's *Human Development Report 2010,* India is home, both as a proportion of the total population and in absolute numbers, to more malnourished children than any other country in the world—more than all the countries of sub-Saharan Africa combined. One-third of the children born in India are underweight, which means that their mothers themselves are seriously undernourished. Each year, millions of infants suffer irreparable cognitive impairment as a result of undernutrition, and millions of children under age two die of hunger. As Sharad Pawar, the Indian minister of agriculture, himself admits, some 150,000 poor farmers committed suicide between 1997 and 2005 to escape the stranglehold of debt. In

2010, more than 11,000 deeply indebted small farmers committed suicide—usually by swallowing pesticides—in the northern states of Bihar, Orissa, Madhya Pradesh, and Uttar Pradesh alone.

In August 2005, in my capacity as UN Special Rapporteur on the Right to Food, I undertook a mission to Shivpur, in Madhya Pradesh, together with my small team of researchers. Shivpur is the name of both a city and its surrounding district, which includes about a thousand villages, each home to between three hundred and two thousand families. In the Shivpur district, the soil is rich and fertile, the forests splendid. But poverty there is extreme and the degree of inequality particularly shocking. Tucked into the valley of the Ganges, Shivpur was, until Indian independence, the summer residence of the maharajas of Gwalior. Of the former splendor of the Shinde royal dynasty there remain today a sumptuous palace in red brick, a polo field, and above all a wildlife park encompassing some 900 square kilometers (about 350 square miles), where peacocks and deer run free. There you can also see a colony of crocodiles living in an artificial lake and caged tigers.

However, the district today remains dominated by a caste of exceptionally voracious great landowners. The district controller, introduced to me as Mrs. Gheeta, age thirty-four, is originally from Kerala; she wears a yellow sari with a narrow red border. I sense immediately that this woman has nothing to do with the bureaucrats we met the day before in Bhopal, the capital of Madhya Pradesh. She is surrounded by department heads, all men with impressive mustaches. Hanging on the wall behind her desk I notice a copy of the famous photograph of Mahatma Gandhi at prayer taken two days before his assassination on January 28, 1948, with, below, these words:

> His legacy is courage,
> His lesson truth,
> His weapon love.

The district controller responds to our questions with extreme discretion, as though she does not trust her mustachioed subordinates. As always, our schedule is tight. We are soon to take some time off.

But first, over the next three days, we will visit the villages and the countryside of the district. We are already expected in Gwalior. And we are already in bed when, late that evening, the hotel receptionist awakens me. A visitor is waiting for me downstairs. It is the district controller of Shivpur. I wake up my colleagues Christophe Golay and Sally-Anne Way. Mrs. Gheeta regales us with a veritable history of her district, ending only at daybreak.

The federal government in New Delhi had demanded that she enforce the new law on agrarian reform and distribute to agricultural day laborers the unused arable land left uncultivated by the great landowners. But as soon as a dalit—one of the so-called untouchables, the poorest of the poor, a member of the most despised castes in India—tries to take possession of his little plot (1 hectare, or about 2.5 acres, per family), he is soon chased off by the great landowners' thugs. The dalit farmers have even been murdered; their killers do not hesitate to eliminate entire families, burn down their huts, and poison their wells. Unsurprisingly, the inquiries opened by the district controller vanish into bureaucratic quicksand. Many of the great landowners maintain useful relationships with the chief minister of Madhya Pradesh in Bhopal, or with the federal ministers in New Delhi. The district controller was close to tears.

In the Indian context, the fight for the legal implementation of the human right to food is of the utmost importance. India inscribed in its constitution the right to life. In its rulings, the Supreme Court of India has asserted that the right to life includes the right to food. Over the course of the last ten years, many rulings have confirmed this interpretation of the law.

Following five years of drought, famine struck the semi-arid state of Rajasthan in 2001. The Food Corporation of India (FCI), a state-owned enterprise that is active throughout entire country, was put in charge of distributing emergency food aid. To this end, the FCI had stockpiled tens of thousands of sacks of wheat in Rajasthani warehouses. But in Rajasthan, as is well known, the representatives of the FCI are especially corrupt. Thus, in order to allow local merchants to sell their wheat at the highest possible price, in 2001 the directors of the FCI refused to distribute their stockpiled grain. The Supreme Court then intervened, ordering the immediate release of the state-owned supplies and the distribution of wheat to starving families. The court's reasoning in support of its verdict is compelling:

> The anxiety of the Court is to see that the poor and the destitute and the weaker sections of the society do not suffer from hunger and starvation. The prevention of the same is one of the prime responsibilities of the Government—whether Central or the State. How this is to be ensured would be a matter of policy which is best left to the Government. All that the Court has to be satisfied and which it may have to ensure is that the food grains which are overflowing in the storage receptacles, especially of FCI godowns [warehouses], and which are in abundance, should not be wasted by dumping into the sea or [being] eaten by the rats. Mere schemes without any implementation are of no use. What is important is that the food must reach the hungry.

Orissa is one of the most corrupt states in India. In the 1970s, the Orissa state government expropriated thousands of hectares of arable land in order to increase the hydroelectric capacity of the Mahanadi River by constructing a series of dams and retention basins. The police accordingly drove thousands of farming families off their land without any compensation whatsoever. The Right to Food Campaign, an Indian NGO, with the support of exceptional lawyers and the leaders of the farmers' unions, filed a complaint with the Supreme Court in New Delhi. The judges ruled against the state of Orissa, requiring that it grant the farmers whose land had been seized "adequate compensation." The court defined what it meant by "adequate compensation" thus: since the rupee had undergone severe inflation in the interim, compensation could not be monetary. The state of Orissa would have to compensate the farmers by providing them with arable land equivalent to their expropriated land in surface area, fertility, soil composition, and access to markets.

As in this case, the Supreme Court of India generally issues extremely detailed verdicts, specifying the exact measures that a state must take to redress such violations of the right to food of which their citizens have been victims. In order to monitor the enforcement of such measures, the court relies on commissioners, specialized officials who are neither judges nor court clerks, but who are bound by oath to uphold the court's rulings. Sometimes these commissioners must work for years to oversee the implementation of programs of reparations and compensation incumbent upon a given Indian state.

It is important to remember that one-third of all the seriously and permanently undernourished people in the world live in India. Farmers whose land is expropriated—for the most part illiterate and the poorest of the poor—obviously have neither enough money nor sufficient familiarity with the justice system to organize themselves to appear before the courts as plaintiffs or to pursue complicated cases against powerful multinational corporations that may last for years, even if they are assisted by state-appointed lawyers. For this reason, the Supreme Court of India allows class action lawsuits. Farmers who are plaintiffs in such suits join together with, for example, civil society organizations, religious communities, and unions that are not themselves plaintiffs in the case. These organizations bring the money, experience, and political muscle needed to embark upon such court battles.

Another legal weapon, unique to the Indian judicial system, allows such organizations to act in this way: public interest litigation. Under this concept, in Indian law, "Any person . . . has the right to appear before a competent court when he maintains that a fundamental right recognized by the constitution has been violated or is threatened with violation. The court may then remedy the situation."

In India, since the right to food is recognized constitutionally, anyone—even people not themselves directly injured—may file a complaint alleging a violation of this right. The legitimacy of such complaints is recognized as inhering in their being in the public interest. In short, every Indian citizen has an interest in ensuring that all human rights, including the right to food, are always and everywhere respected by every level of government. Being based on this concept of the public interest, such complaints are of enormous practical importance. In such states as Bihar, Orissa, or Madhya Pradesh, members of the upper castes have a monopoly on power, controlling practically all administrative and judicial functions. Many are corrupt to the marrow. Toward the dalits and the indigenous forest-dwelling peoples, the upper castes express limitless contempt. Government ministers, police officers, and local judges terrorize farmers whose land has been confiscated.

Colin Gonsalves, founding executive director of the Human Rights Law Network, one of the principal directors of the Indian Right to Food Campaign, and an advocate who appears before the Indian supreme court, has recounted the extreme difficulty advocates face, under such conditions, in persuading the heads of families who have been illegally deprived of their huts, their wells, and their plots of land to file a complaint and appear before a local judge. The farmers tremble in fear of the Brahmins. Nonetheless, public interest litigation does make it possible to take an Indian state to court if it confiscates farmers' lands without their consent. It is in the state of Madhya Pradesh that the Supreme Court is most active.

In India, some eleven thousand farming families were evicted from their land in 2000 by local governments to make way for hydroelectric dams or the development of mines. In Hazaribagh, in Jharkhand state, thousands of families have had their land expropriated by the state and turned over to coal mining. The construction of the gigantic Sardar Sarovar dam on the Narmada River in Gujarat has deprived thousands of families of their means of subsistence. Their complaints for redress and compensation in kind are currently before the courts.

Recalling the countryside of Madhya Pradesh reminds me of unforgettable images of skeletal children with enormous eyes, "astonished to suffer so much," as Edmond Kaiser (the founder, in 1960, of Terre des Hommes) used to say, with bitter irony. For the people of Madhya Pradesh (one of the most impoverished states of India), so hospitable, so warm, the daily quest for a handful of rice, an onion, a piece of bread requires all their energy.

The second obligation imposed on signatory states by the UN's

Covenant on Economic, Social and Cultural Rights is this: the state itself must not only "respect" the right to food of its inhabitants but must also "protect" this right against violations inflicted by third parties. If a third party infringes upon the right to food, the state must intervene to protect its inhabitants and to reestablish the violated right.

Consider the example of South Africa. There, the right to food, inscribed in the constitution, enjoys extended protection. In South Africa there is a national Human Rights Commission, comprising equally representatives of the government and civil society organizations (unions, churches, women's movements, and so on). This commission may seek redress from the Constitutional Court in Pretoria and from the South African regional high courts for any law enacted by Parliament, any governmental regulation, any decision taken by a government official, or any action undertaken by a private enterprise that violates the right to food of any group of citizens. South African jurisprudence is in this regard exemplary.

The right to drinking water is also encompassed within the right to food. A few years ago, when the city of Johannesburg engaged a multinational corporation to provide its drinking water, the company massively raised the rates for water, to exploitative levels. Many inhabitants of the city's poor neighborhoods, unable to pay such exorbitant rates, had their supply of running water cut off. Since the company also required prepayment for water usage in excess of 25 liters (6.6 gallons), many families of modest means were reduced to fetching their water from drains, gutters, ditches, polluted streams, and ponds. Supported by the Human Rights Commission, five inhabitants of the township of Phiri, in Soweto, took their case to the high court.

And they won. The city of Johannesburg was required to reestablish its former system of providing water at low prices as a public service.

Article 11 of the Covenant on Economic, Social and Cultural Rights stipulates a third obligation of every signatory state: when famine strikes, if the government concerned is not capable of combating the disaster with the means at its disposal, it must appeal for international aid. If it does not do so, or does so only after a deliberate delay, the country violates its inhabitants' right to food.

In 2006, a terrible famine, caused by both locusts and drought, struck south and central Niger. Many grain merchants flatly refused to put their stockpiled grain on the market. They waited for the shortages to worsen and for prices to rise. In July 2005, I thus found myself on a mission to Niger, and in the office of the country's president, Mamadou Tandja. He denied the evidence. Apparently he was in collusion with the speculating merchants.

It took CNN, Doctors Without Borders, and the NGO Action Against Hunger arousing worldwide public outrage, and a personal three-day visit from Kofi Annan to Niger's Maradi and Zinder regions, to force Niger's government to finally make a formal appeal for aid to the World Food Programme. Meanwhile, tens of thousands of men, women, and children were already dead by the time the first truckloads of international aid—sacks of rice and flour and containers of water—arrived in Niamey.

Tandja, apparently, was at no point concerned, since the survivors of the famine had no means to accuse him in court.

For the WTO, the U.S. government (as well as the Australian, British, and Canadian governments, among others), the IMF, and the World Bank, all these normative interventions provided for in the despised covenant are anathema. In the eyes of the supporters of the "consensus of Washington," they constitute an intolerable attack on the free market.

Those who are called in the South the *corbeaux noirs du FMI* (the "black ravens of the IMF"; or in English, the "IMF vultures") consider the arguments advanced by supporters of the right to food to be pure ideology, doctrinaire blindness, or, worse, communist dogma.

There is a cartoon by the French cartoonist Plantu that shows an African child dressed in rags standing behind a hugely fat white man wearing glasses and a tie and seated before a sumptuous meal. The child says, "I'm hungry." The fat white man turns to him and replies, "Stop talking politics!"

14

THE HORSEMEN OF THE APOCALYPSE

The three horsemen of the apocalypse of organized hunger are the WTO, the IMF, and, to a lesser extent, the World Bank. The World Bank was directed until recently by Robert Zoellnick, George W. Bush's former chief trade negotiator; the IMF is currently directed by Christine Lagarde and the WTO by Pascal Lamy. These three have in common exceptional expertise, intellectual brilliance, and an impenetrable faith in neoliberalism. One curious detail: Lamy is a member of the Parti Socialist Français. All three directors are top-flight technocrats and ruthless, remorseless realists. Together, all three exercised exceptional power over the economies of the most fragile countries on the planet, and Lagarde and Lamy continue to do so. Contrary to the recommendations of the UN Charter, which assigned this task to the Economic and Social Council, it is the directors of these three agencies who determine the policies of the UN.

The IMF and the World Bank were founded in 1944 in the tiny resort town of Bretton Woods, New Hampshire. Both are integral parts of the UN system. The WTO, on the other hand, is a totally autonomous organization, independent of the UN, although it comprises representatives of 157 countries (as of August 24, 2012) and functions by negotiated consensus. The WTO was founded in 1995 as the successor to the General Agreement on

Tariffs and Trade (GATT), which had been established by the industrial nations in the aftermath of World War II to harmonize and gradually lower customs tariffs.

René Dumont's successor as a leading theorist of world agriculture, Marcel Mazoyer, is professor emeritus of comparative agriculture and agricultural development at the Institut National Agronomique de Paris-Grignon (the French National Agricultural Institute, popularly known as Agro-Paris-Tech). In 2009, before an audience of UN–accredited ambassadors in Geneva, he subjected the WTO's policies to a merciless critique: "The liberalization of agricultural exchanges, by reinforcing competition among extremely unequal forms of agriculture, as well as price instability, can only aggravate the food crisis, the economic crisis, and the financial crisis."

What is the goal that the WTO pursues when it fights in favor of the total liberalization of trade, patents, capital, and financial services? Rubens Ricupero, the former secretary general of the United Nations Conference on Trade and Development (UNCTAD) from 1995 to 2004 and former finance minister of Brazil, gives a straightforward answer to that question: "The unilateral disarmament of the countries of the South."

The IMF and the WTO have always been the most determined enemies of economic, social, and cultural rights, and particularly of the right to food. The 2,000 IMF bureaucrats and the 750 more at the WTO regard with horror any normative intervention in the free play of the market, as I have said. The organizations' policies have not fundamentally changed since their founding, even if Dominique Strauss-Kahn, the IMF's director from 2007 until his resignation in 2011, did allow a larger role for emerging nations in the IMF's governance, and even if he did make an effort to develop a more favorable lending policy for poor countries—which are regardless sooner or later always reduced to bankruptcy.

A simple image enables us to grasp the justice of Mazoyer's and Ricupero's views. Imagine that Mike Tyson, former world heavyweight boxing champion, were to confront in the ring an undernourished, unemployed Bengali worker. What would the ayatollahs of neoliberal dogma say? That a fair fight is ensured, because both fighters have to wear the same kind of boxing gloves, fight the same number of rounds in the same space, and abide by the same rules. So, may the best man win! The impartial judge is the market. The absurdity of neoliberal dogma stares you right in the face.

During my two terms as UN Special Rapporteur on Right to Food, I got to know four successive American ambassadors to the UN's agencies in Geneva. All four, without exception, fought vigorously against my reports and all my recommendations. Twice they demanded (unsuccessfully) that Kofi Annan fire me, and, of course, they voted against the renewal of my appointment.

Two of these ambassadors—and one in particular, a nabob of the pharmaceutical industry from Arizona, a special envoy of George W. Bush's—expressed their personal hatred toward me. Another contented himself with strictly applying State Department directives: the refusal to recognize the existence of economic, social, and cultural rights and to recognize only civil and political rights.

With one of the four I did develop a friendly relationship. George Moose was President Bill Clinton's ambassador, a cultivated African American man with a subtle mind, who was accompanied by his wife, Judith, an obviously left-leaning intellectual, warm and funny, who also worked in the State Department. Before his appointment to Geneva, George Moose had held the post of assistant secretary of state for African affairs. It was he who, in 1996, had chosen Laurent Kabila—then an obscure guerilla fighter and gold trafficker hiding out in the mountains of Maniema—as leader of the Alliance of Democratic Forces for the Liberation of Congo-Zaire (AFDL), now the Democratic Republic of Congo. With his passion for history, Moose knew that Kabila was the only surviving leader of the pro-Lumumba rebellion against Mobutu Sese Seko in 1964 who had never sworn allegiance to Mobutu and who retained intact his credibility with the youth of Congo. Events would bear out the wisdom of Moose's choice.

But our shared passion for Africa was not enough. Throughout his tenure in Geneva, Moose, like the other ambassadors, fought against every one of my recommendations and initiatives, every one of my reports on the right to food. I was never able to pierce the armor of his genuine convictions in this regard.

For more than two decades, privatization and the liberalization of trade, financial services, capital flow, and patents have proceeded at a stupefying pace. As a result, the poor countries of the South have found themselves largely stripped of the prerogatives of sovereignty. Borders have disappeared, and the public sector—even hospitals and schools—has been privatized. And all over the world, the number of victims of undernutrition and hunger grows.

One study by Oxfam that has become famous showed that in every country where the IMF has instituted a program of "structural adjustment" in the decade from 1990 to 2000, millions more human beings have been thrown into the abyss of hunger. The reason is simple: the IMF is, precisely, in charge of the management of the external debt of 122 countries of the so-called Third World, which, as of December 31, 2010, totaled $2.1 trillion. In order to make the interest and capital payments on its debts held by creditor banks or the IMF itself, a debtor country needs foreign currency. The big creditor banks obviously refuse to be repaid in Haitian gourdes, Bolivian bolivianos, or Mongolian tugriks. How can a poor country in South Asia, South America, or sub-Saharan Africa acquire the necessary foreign currency? By exporting manufactured goods or raw materials that will be paid for in foreign currency. Of the fifty-three countries in Africa, thirty-seven have almost entirely agricultural economies.

The IMF periodically grants overindebted countries a temporary moratorium on their debt repayments or a refinancing of their debt—on the condition that the overindebted country submit to a program of so-called structural adjustment. All these plans require the reduction of expenditures on health and education in the budgets of the country involved, and the elimination of subsidies for staple foods and aid to families in need. Public services are the first victims of structural adjustment: thousands of public service employees, including nurses and teachers, have thus lost their jobs in countries all over the world.

In Niger, as I have said, the IMF demanded the privatization of the National Veterinary Office. Ever since, Niger's stockbreeders have had to pay exorbitant prices set by multinational corporations for the vaccines, vitamin supplements, and antiparasitic medications they need for their livestock. The consequence? Tens of thousands of families have lost their herds. Today these people waste away in the shantytowns of the big coastal cities of West Africa: Cotonou, Dakar, Lomé, Abidjan.

Wherever the IMF holds sway, the fields of manioc, rice, or millet shrink. Subsistence agriculture dies. The IMF requires the expansion of the crops of colonial agriculture, whose products—cotton, peanuts, coffee, tea, cocoa—can be exported and sold on the world market to bring in foreign exchange, which will in turn go to service the national debt.

The second task of the IMF is to open up markets in the countries of the South to private global food corporations. This is why, in the southern hemisphere, free trade wears the hideous mask of famine and death. Consider the following examples.

Haiti is today the poorest country in Latin America and the third-poorest country in the world. In Haiti, rice is the staple food. In the early 1980s, Haiti was self-sufficient in rice. Working terraced fields and the wet lowlands, Haitian farmers were protected from foreign dumping by an invisible wall: a tariff of 30 percent on imported rice. But over the course of the 1980s, Haiti was subjected to two programs of structural adjustment. Under orders from the IMF, the protective tariff was reduced from 30 to 3 percent. American rice, which is heavily subsidized by the U.S. government, flooded into Haitian towns and villages, destroying the country's rice production and, as a consequence, the way of life of tens of thousands of rice farmers. Between 1985 and 2004, Haitian imports of foreign rice, mainly American and heavily government-subsidized, increased from 15,000 tons to 350,000 tons annually. At the same time, local rice production collapsed, declining from 124,000 tons to 73,000 tons.

Since 2000, the Haitian government has had to spend more than 80 percent of its meager revenues to pay for imported food. And the destruction of rice farming has caused a massive exodus from the countryside. The overcrowding of Port-au-Prince and the country's other big cities has led to the disintegration of public services. In short, Haitian society has been totally turned upside down, weakened, made more vulnerable than ever before by the effects of the IMF's neoliberal policies. And Haiti has been reduced to a beggar state, subject to foreign laws. Coups d'état and social crises have followed inevitably, one after another, throughout the last twenty years.

Normally, the 9 million people of Haiti consume 320,000 tons of rice per year. When world prices for rice tripled in 2008, the Haitian government was unable to import enough food. In Cité Soleil, between the hill that dominates Port-au-Prince and the Caribbean—one of the largest shantytowns in Latin America—hunger began to prowl the streets.

Since the 1990s, Zambia has endured a whole series of structural adjustment programs. The social and nutritional consequences for the population have been, obviously, catastrophic.

Zambia is a magnificent country, where the Zambezi River flows gently and the farmers' sloping fields grow lush and green thanks to the mild climate. The staple food of Zambia's people is corn. At the beginning of the 1980s, the price of Zambian corn was 70 percent subsidized by the state for consumers; producers were also subsidized. Sales of corn on the domestic market and exports to Europe (in years of surplus) were regulated by the state's marketing board. Together, subsidies to consumers and producers absorbed slightly more than 20 percent of the federal budget. Everyone had enough to eat.

The IMF imposed first the reduction and then the abolition of the subsidies. It also forced the state to eliminate subsidies for the purchase of fertilizer, seed, and pesticides. Schools and hospitals, which had previously been free, began to charge fees. With what consequences? In the countryside and in poor urban neighborhoods, families were reduced to eating no more than one meal per day. Subsistence agriculture began to decline and then collapse as farmers were deprived of fertilizer and improved seed. In order to survive, they sold their draft animals, which led to further declines in productivity. Many were forced to leave their land and sell their labor as agricultural day laborers on the big cotton plantations owned by foreign corporations. Between 1990 and 1997, the consumption of corn fell by 25 percent. The result: the rate of child mortality exploded. By 2010, 86 percent of the Zambian population lived below the national poverty line. In the same year, 72.6 percent of the population lived on less than $1 a day; 45 percent of Zambians were seriously and permanently malnourished. Forty-two percent of children under five years old were 24 percent below the normal weight for their age as defined by UNICEF.

The American mentality dominates in the glass-clad building at 700 19th Street NW in Washington, D.C., the headquarters of the IMF. The Fund's annual reports reveal a delightful candor. A report for 1998, for example, admits of one program of structural adjustment that while in the long term the program will improve access to resources and increase the population's income, in the short term it is reducing the consumption of food.

At the level of the Zambian state itself, the program of structural adjustment has had disastrous consequences. The tariffs protecting national industries were abolished, and most of the public sector was privatized. The revision of the Employment Act and the Land Act badly damaged the social safety net, union rights, and the right to a guaranteed minimum salary. There followed mass evictions of people from their homes, mass unemployment, and a drastic rise in the price of staple foods.

The IMF bureaucrats do have a sense of humor. In the conclusion to their report, they applaud the fact that disparity of living conditions between Zambia's urban and rural populations was considerably reduced between 1991 and 1997. Why? Because the level of urban poverty increased so dramatically that it equaled the level of poverty in the countryside.

Aside from Ethiopia and Liberia, Ghana was the first country in Africa to have achieved its independence. After repeated general strikes, mass uprisings, and ferocious repression by the British colonial regime, the Republic of Ghana, the successor to the mythic kingdom of the Kaya Maga or Kanne Mahan (a name that, in the language of the Soninke people, the founders of the ancient kingdom of Ghana, means "king of gold"), was born in 1957. Its flag depicts a black star on a white ground. Ghana's first president, Kwame Nkrumah, the prophet of pan-African unification, was one of the founders, together with Gamal Abdel Nasser, Modibo Keita, and Ahmed Ben Bella, of the Organization of African Unity in Addis Ababa in 1960.

Ghanaians of all the country's fifty-two ethnic groups are fiercely proud people, viscerally devoted to their national sovereignty. Nonetheless, they too had to bow before the IMF and the multinational food corporations. Ghana has suffered a fate similar in every detail to Zambia's.

In 1970, some 800,000 local farmers supplied all the rice consumed in Ghana. In 1980, the IMF struck for the first time: the protective tariff on rice was reduced to 20 percent, then lowered further still. The IMF then demanded that the state eliminate all subsidies to farmers for the purchase of pesticides, mineral fertilizers, and seed. Today, Ghana imports more than 70 percent of the rice consumed in the country. Ghana's Marketing Board, the national office that supported the commercialization of agricultural products, was abolished. Private companies have since taken over agricultural exports.

Ghana is a lively democracy whose parliamentary representatives are animated by a strong sense of national pride. In order to revive national rice production, the parliament in Accra decided in 2003 to introduce a 35 percent tariff on imported rice. The IMF reacted forcefully and required the government to revoke the law.

In 2010, Ghana spent more than $400 million on food imports.

In the same year, Africa as a whole spent $24 billion to finance its importation of food. As I write, in 2011, speculation on agricultural commodities exchanges has caused a worldwide explosion of the prices of staple foods. In all likelihood, Africa will not be able this year to import enough food.

Everywhere and always, the violence of the free market and the arbitrariness of its operations, free from all normative constraint, from any kind of social control, kills—through poverty and through hunger.

15

WHEN FREE TRADE KILLS

In December 2005, during a WTO Ministerial Conference in Hong Kong that aimed to relaunch the negotiation process begun in Doha in 2001, and which has been stalled ever since, the WTO attacked free food aid. The WTO declared that it was unacceptable that the WFP and other organizations distribute—in refugee camps, villages ravaged by locusts, hospitals where undernourished children lie dying—rice, precooked cereal-and-pulse blends, high-energy biscuits, or milk for free, taking advantage of the surplus agricultural products provided to the WFP by donor nations. According to the WTO, this practice corrupts the market. Any commercial transfer of a good must have a price. The WTO therefore demanded that in-kind aid provided to the WFP by donor nations be henceforward taxed at its fair value. In short, the WFP must no longer accept in-kind donations originating as surplus agricultural production in donor nations, and must no longer distribute anything except food purchased on the open market.

Thanks in particular to Daly Belgasmi, former director of the Geneva office of the WFP (and currently WFP regional director for the Middle East, Central Asia, and Eastern Europe), and to Jean-Jacques Graisse, the WFP's senior deputy executive director and chief of operations, the WFP reacted forcefully in a memorandum:

A woman widowed by AIDS in Zambia with six small children is not concerned to know whether the food aid she receives comes from an in-kind donation to the WFP or from
a monetary contribution by the same donor. All that she
wants is for her children to live and not to have to beg for
food. . . . WHO tells us that, on our Earth, undernutrition
and hunger constitute the most important risks to human
health. Every year, hunger kills more human beings than
AIDS, tuberculosis, malaria, and all other epidemic diseases combined. . . . The WTO is a club of the rich. . . .

The debate that the WTO is initiating is not a debate
about hunger, but a debate about commercial advantages. . . . Is it tolerable to reduce food aid for starving mothers and children who play no role in the world market in the
name of economic liberalism?

The memorandum concluded: "We want world trade to be endowed with a conscience."

In Hong Kong, the countries of the southern hemisphere stood
up to the powerful interests that dominate the WTO. The propo-
sition to tax food aid was rejected. Pascal Lamy, the WTO director general, and his associates were soundly defeated.

The WTO soon suffered another defeat, this time at the hands
of India. The jurisprudence of the Supreme Court of India is beyond the reach of the WTO. India, to be sure, is a member of
the WTO, but the organization's statutes impose obligations only
upon the executive power of its member states, not on their judi-
ciaries. India is a large and vibrant democracy; its system of government relies upon the separation of powers.

However, India's Public Distribution System (PDS) exercises
executive power. What does this mean? In 1943, a terrible famine left more than 3 million dead in Bengal. The English colonial
government had emptied the granaries, requisitioning the harvests and sending the confiscated food to the British forces fighting the Japanese in Burma and on the other Asian fronts. After
the Bengal famine, Mahatma Gandhi made the fight against hun-

ger an absolute priority of his struggle. Jawaharlal Nehru, the first prime minister of independent India, continued the fight. Today, if even one person dies of hunger in any of the country's six thousand districts, the district controller is immediately dismissed.

This reminds me of a night in August 2005 in Bhubaneswar, the magnificent capital city of the state of Orissa, on the coast of the Gulf of Bengal. Each of my missions included as a top priority meetings with representatives of social movements, religious communities, unions, and women's movements. In Bhubaneswar, Pravesh Sharma, on behalf of the Indian branch of the International Fund for Agricultural Development, which has its headquarters in Rome, was in charge of organizing these meetings. More than 40 percent of Indian farmers are landless peasants, tenant farmers, and migrant workers who travel from harvest to harvest. IFAD works above all with tenant farmers, who live in abysmal poverty. Sharma introduced us to two women in faded brown saris, sad-faced but determined; each had lost a child to hunger. My colleagues and I listened at length, taking notes and asking questions. The meeting took place out in the suburbs, far from our hotel and from the local UN offices.

Three days later, in the departure lounge at the Bhubaneswar airport, a police officer approached me. A delegation, dispatched by the chief minister of Orissa, was waiting for me in a meeting room. The group was led by P. K. Mohapatra, the senior regional manager of the Food Corporation of India in Bhubaneswar. For three hours, the five men and three women of the delegation tried to persuade me, with the help of medical certificates and other documents, that the two children I had heard about three days before had died not from hunger but from an infection. Evidently there was a great deal at stake for some of these officials—their jobs were on the line.

The FCI administers the Public Distribution System, maintaining immense warehouses in each Indian state, buying wheat in Punjab and stockpiling it in every corner of the country. Across the land, the FCI oversees more then half a million storehouses. Local assemblies in villages and urban neighborhoods draw up lists of recipients of food aid. Every family that receives aid gets a card confirming their eligibility. There are three categories of recipients of PDS food aid: those Above the Poverty Line (APL); those Below the Poverty Line (BPL); and those who qualify for the Antyodaya Anna Yojana (AAY) scheme, which seeks to assist the poorest of the poor. For each of these three categories, there are specific set prices for each kind of food available. A family of six has the right to 35 kilograms of wheat and 30 kilograms of rice per month. In 2005, for a BPL family, the prices were as follows: 5 rupees for a kilogram of onions, 7 rupees for a kilogram of potatoes, and 10 rupees for a kilogram of cereal (rice, wheat, corn). (In 2005, the rupee traded at between 2.2 and 2.3 cents to the U.S. dollar.) It is important to understand that the minimum wage in India for urban workers in 2005 was 58 rupees per day.

It is true that 20 percent of PDS stocks of food are regularly sold illegally on the open market. Some ministers and officials make fortunes by misappropriating stockpiled food supplies in this way. Corruption is endemic. Nonetheless, hundreds of millions of extremely poor people benefit from the PDS. Since the "farmers' prices" in the FCI food ration shops are a small fraction of market prices (varying somewhat, as noted above, according to the category of the recipients), large-scale famine has been eradicated in India.

Moreover, the PDS system improves children's lives. There are in fact more than 900,000 specialized facilities for child nutrition in India, the Integrated Child Development Centers (ICDs). According to UNICEF, more than 40 million of the 160 million children in India under age five are seriously and permanently undernourished. For some of them, the ICDs provide therapeutic feeding, vaccines, and sanitary care. The ICDs are supplied by the FCI. In the fight against the scourge of hunger, the PDS thus plays a crucial role. If the WTO undertook to eliminate the PDS, it is because the system's very existence and its manner of functioning are in effect contrary to the statues of the WTO.

Hardeep Singh Puri, the current permanent representative of India to the United Nations, who is a Sikh and a man of inexhaustible energy, has fought very hard against the WTO's attempt to abolish the PDS. In New Delhi he has relied on two allies as determined as he is: his brother, Manjeev Singh Puri, who was from 2005 to 2009 the joint secretary (United Nations–Economic and Social) in the Indian Ministry of External Affairs, and who is now India's deputy permanent representative to the United Nations; and Sharad Pawar, the minister of agriculture. Together, these three men have saved the PDS and outwitted the WTO.

16

SAVONAROLA ON LAKE GENEVA

Pascal Lamy, the director general of the WTO, is the Savonarola of free trade. The man's willpower and analytical intelligence are impressive. His current position and his previous career give him influence and prestige enjoyed by few other directors of international organizations today: as I have noted, the WTO currently has 157 member states, and its headquarters on Rue de Lausanne in Geneva employs 750 officials.

Lamy is an austere man, even ascetic; he runs marathons. According to Lamy himself, he travels nearly 450,000 kilometers (300,000 miles) each year by plane, apparently immune to the havoc wrecked on the human body by jet lag—not to mention the interminable nighttime meetings that are customary at the WTO.

Lamy is not given to soul-searching. To one interviewer, he declared, "I am neither an optimist nor a pessimist. I'm an activist." Lamy is a man of power; power relations alone interest him. When the same interviewer remarked, "Like the IMF, you are publicly accused by some of driving the poorest citizens of the poor countries deeper into poverty," the director of the WTO replied, "An agreement always reflects the power relations in force at the moment when it is signed."

A former European commissioner for trade, Lamy has shaped

the WTO since its earliest days. One of his books, *L'Europe en pre-mière ligne*, gives an account of his indefatigable war against all forms of normative (rights-based) or social control of markets.

Member nations' ambassadors and delegates to the WTO, as well as Lamy's colleagues, are obviously fascinated by him. And like Savonarola in fifteenth-century Florence, nothing escapes Lamy. He is constantly on alert, pitilessly hunting down deviants from free-trade dogma. His informants are everywhere.

I myself once experienced this. Every September, an extraordinary man named Jean-François Noblet hosts a Festival of Life in the little town of Albenc, in the mountains of the Dauphiné, a few dozen miles from Grenoble, bringing together representatives of the region's social movements, unions, and religious communities. I was invited to speak there in 2009. In my address, I criticized, in measured fashion, the WTO's policies regarding trade in food. A full moon shone in the sky. The big tent was full of people. The impassioned discussion that ensued continued past midnight.

But one of Lamy's men (or women) was in the audience. On September 29, 2009, I received the following letter from him:

> Dear Jean,
>
> To my consternation, I have been made aware, once again, of remarks that you have made, this time during a conference in Albenc, casting me in a defamatory light: my actions are, according to you, "totally contrary to the interests of the victims of famine." On the contrary! The WTO is eager to conclude the Doha Round . . . which will be equivalent to killing more people . . . ? [. . .]
>
> Obviously absurd! The members of the WTO have been negotiating for eight years the mandate that they have given themselves, at the request of the developing countries, to open up agricultural markets further, and above all those of the developed countries, to which they [the developing countries] want to be able to have access. [. . .]
>
> The simplest thing, to give you some idea of the reality of the situation, would be for you to ask the representatives

of these countries what they think. This is, by the way, what
your successor, Olivier De Schutter, did, in the course of
a discussion during a WTO Agriculture Committee meeting, whose outcome left little doubt about the position of the
countries in question. [. . .]

With the hope that this reminder of a few political realities will prevent you in future from making such false accusations, my dear Jean, I remain,

Yours, etc.

Obviously I didn't need anyone to suggest that I "consult" representatives of the governments of the South. In fulfilling my duties, I met with them daily. Some are my friends. But Lamy is right
on one point: few among them openly oppose the WTO's policies
regarding agricultural trade. The reason is obvious: many governments in the southern hemisphere depend for their very survival
on development aid, capital investments, and infrastructure credits
from the Western countries. Without regular remittances from the
European Development Fund (EDF), for example, many governments in sub-Saharan Africa, the Caribbean, and Central America
would be unable to pay the salaries of their government ministers,
their bureaucrats, and their soldiers year-round. The WTO is a
club of the rich and dominant nations. This fact inspires prudence.

Lamy mentions the opening of markets in the industrialized
countries to the products of Southern farmers. In this he sees
proof of the willingness of the WTO to come to the aid of Third
World farmers. But this "proof" is unconvincing.

During the WTO Ministerial Conference in Cancún in 2003,
an international accord on agriculture was supposed to be formal-
ized that would, among other things, open agricultural markets in
the South to the multinational food industry corporations of the
North in return for providing access to the markets in the North
to certain products from the South. In Cancún, the Brazilian ambassador, Luiz Felipe de Seixas Corrêa, organized resistance to
the accord. The countries of the South refused to open their markets to global private corporations and foreign sovereign wealth

funds (state-owned investment funds). Cancún was a complete fiasco. And to this day, the international accord on agriculture—the centerpiece of the Doha Round—has not been signed. For everyone in the South knows that the prospect Lamy holds out of opening Northern agricultural markets to Southern products is just an illusion. (One reservation, however: for the fifty least developed countries, certain products are permitted exceptional access to Northern markets.)

In the philippic that he sent me, Lamy talks about the elimination of export subsidies that rich countries pay their farmers. The WTO's 2005 Hong Kong Ministerial Declaration states: "We agree to ensure the parallel elimination of all forms of export subsidies and disciplines on all export measures with equivalent effect to be completed by the end of 2013. This will be achieved in a progressive and parallel manner."

The problem is that negotiations leading to the elimination of export subsidies have never proceeded beyond the stage of declarations of intention. Negotiations with a view to an international accord on agriculture are at a standstill. And the rich countries continue to subsidize their farmers massively. In any African market, in Dakar, Ouagadougou, Niamey, or Bamako, a housewife can buy vegetables, fruits, or chickens from France, Belgium, Germany, Spain, Greece . . . at one-half or one-third the price of equivalent African products. A few miles away, Wolof, Bambara, or Mossi farmers, their wives, and their children exhaust themselves working twelve hours a day under the blazing sun without the least chance of ensuring for themselves the bare subsistence minimum.

As for Olivier De Schutter, my outstanding successor, Lamy has no doubt not read the report he wrote following his mission to the WTO. This report principally discusses the international accord on agriculture that the WTO has failed to conclude since the setback at the Cancún conference in 2003. In it, De Schutter severely criticizes the WTO's policies, writing:

> The report argues that, if trade is to work for development and to contribute to the realization of the right to food, it

needs to recognize the specificity of agricultural products, rather than to treat them as any other commodities; and to allow more flexibilities to developing countries, particularly in order to shield their agricultural producers from the competition from industrialized countries' farmers.

Virtually all NGOs and farmers' unions, as well as the governments of many countries in the South, have demanded that the accord on trade in agricultural goods be excluded from WTO oversight and removed from the Doha Round. Food, they say, should be considered as a public good. De Schutter has come round to this view.

Rereading Lamy's letter, I think of the man of exceptional gifts, but proud, whom Ecclesiastes addresses: "The lord will smite you with madness and blindness and confusion of mind."

PART IV

THE COLLAPSE OF THE WFP AND THE FAO'S IMPOTENCE

17

A BILLIONAIRE'S FEAR

The Food and Agriculture Organization and the World Food Programme are the great and beautiful legacy of Josué de Castro. Both were born, as I have described, out of the tremendous awakening of conscience that seized Europe in the aftermath of fascism, the FAO in 1946 and the WFP in 1963. Today, these two institutions lie in ruins.

The WFP enjoys less sumptuous offices than the FAO. Its world headquarters is located in a rather drab suburb of Rome, adjacent to a cemetery, some vacant lots, and a ceramics factory. Nonetheless, the WFP is the most powerful humanitarian organization in the world—and one of the most effective. The WFP's mission is to provide emergency humanitarian aid. In 2010, the list of recipients of WFP aid included nearly 90 million hungry men, women, and children. The WFP currently employs slightly more than ten thousand people, 90 percent of whom work in the field with the victims of hunger.

Within the UN system, the WFP enjoys great independence. It is directed by an administrative council comprising representatives of thirty-six UN member states. One member state, the United States, provides about 60 percent of the contributions to the WFP. Over the decades, American contributions have been above all in kind; the United States has dumped its enormous agricultural surpluses onto the WFP. However, times have changed. The American surpluses are melting away very rapidly, especially owing to the large-scale manufacture of biofuels, which is subsidized by billions of dollars in public funds. This is why, since 2005, in-kind contributions provided by Washington to the WFP have fallen by 80 percent. Yet the United States remains, by far, the leading contributor to the WFP in terms of monetary donations. European support is more reduced: in 2006, the UK donated $835 million and Germany $340 million, while France's contribution remains very low, $67 million in 2005 and $82 million in 2006.

In order to reduce transportation costs as much as possible, but also to avoid penalizing Southern farmers, the WFP tries very hard to buy food in countries as close as possible to hunger zones. In 2010, the WFP spent some $1.5 billion on food. In 2009–10, WFP aid was devoted first and foremost to three specific populations: the victims of the floods in Pakistan, the drought in the Sahel, and the earthquake in Haiti. In 2010 as well, thousands of tons of corn, rice, wheat, and special foods for children under age two and for pregnant and nursing mothers were purchased in Argentina, Mexico, and Thailand, but also in Europe (mainly intravenous therapeutic nutrition products). On February 11, 2011, at a press conference held in Rome, Josette Sheeran, then the WFP's executive director, was able to confirm that in 2010, for the first time, the WFP bought more than 80 percent of its food in the southern hemisphere.

In the first chapter of this book, I outlined the clear distinction that the UN makes between structural hunger, which it is the FAO's mission to combat, and conjunctural hunger, which the WFP tries to reduce. This distinction must be made more nuanced in the context of the WFP's work.

According to its mission statement, the mandate of the WFP is very precise:

> The policies governing the use of World Food Programme food aid must be oriented towards the objective of eradicating

hunger and poverty. The ultimate objective of food aid should be the elimination of the need for food aid. . . . Consistent with its mandate, which also reflects the principle of universality, WFP will continue to: use food aid to support economic and social development; meet refugee and other emergency food needs, and the associated logistics support; and promote world food security in accordance with the recommendations of the United Nations and FAO.

According to its mandate as originally defined by the UN General Assembly, the WFP must in particular reduce the toll of child mortality, improve the health of pregnant women, and fight against micronutrient deficiencies. This is why, beyond emergency food aid, the WFP provided, until 2009, school meals for 22 million children living in the poorest countries. However, most of these meals have recently been eliminated, for reasons that I discuss below.

The WFP was also the pioneer of a method of emergency intervention called Food for Assets (also called Food for Work), according to which victims of hunger who are able-bodied enough are hired by the WFP to work on the repair and reconstruction of damaged roads, bridges, irrigation canals, silos, schools, and hospitals, and on soil rehabilitation projects. In exchange for their labor, men and women who are heads of families are paid in kind: so many sacks of rice for so many days of work. Moreover, all the Food for Assets projects are conceived by local people themselves, and they are the ones who decide which projects will be started first.

The first time I saw a Work for Assets project under way, I was in the southern Caucasus, in Georgia. This magnificent and very ancient country has recently been torn apart by two civil wars. Immediately following the dismantling of the USSR in 1991, South Ossetia and Abkhazia, two separatist regions of Georgia, declared independence (in 1992 and 1993, respectively). The Tbilisi government tried to crush the rebels. To escape the fighting, tens of thousands of refugees, including members of the

Georgian majorities of both regions, fled into Georgia. Given the stagnation that followed the collapse of the Soviet economy, Georgia did not have the means to feed and care for the refugees. The WFP took on the task, trying to do the least possible harm.

The two autonomous regions were ravaged. In both, the WFP financed the clearing and rehabilitation of tea plantations abandoned by farmers fleeing the fighting. In Georgia, refugee farmers were put to work by the WFP on large-scale construction projects and were paid not in cash but in sacks of rice, wheat, and powdered milk. Thanks to the WFP, for the last two decades, thousands of persecuted families, victims of the vast waves of "ethnic cleansing" that took place in the course of these two civil wars, have been able to feed themselves almost normally.

Since that time, I have seen Work for Assets in action on the arid plateau near Mek'ele, the capital of the Tigray region in northern Ethiopia, where nothing grows in the stony ground except a few wretched stalks of teff; also on the Yucatán sierra; in Guatemala; and on the steppe in Mongolia's Selenge province, on the edge of the vast Siberian taiga. Everywhere, I have been impressed by the eagerness with which entire families have joined in the program. Work for Assets transforms victims into makers of their own future, restores their dignity, helps to reconstruct badly damaged societies, and, in the WFP's own words (which I saw on a sign at a WFP jobsite in Rajshahi, Bangladesh), helps to turn hunger into hope.

The WFP also leads exemplary diplomatic campaigns. Like the International Committee of the Red Cross (ICRC), the WFP expresses doubts about the effectiveness of so-called humanitarian corridors, the supposedly "neutral" zones that the UN seeks to create so that food can be transported from central depots to camps for displaced persons whom the UN is attempting to aid. The idea is, however, attractive: in open war, don't humanitarian corridors guarantee the free access of trucks bringing aid? Yet establishing the humanitarian corridors suggests to the warring parties that outside the corridors' perimeter, everything is permitted, including poisoning wells and fields, slaughtering livestock,

burning harvested food, destroying crops—all this, in defiance of
the Geneva Conventions and other international norms governing the protection of civilians and the environment in wartime.

In western Sudan, in northern Kenya, in western Pakistan, in
Afghanistan, armed gangs or soldiers periodically attack WFP
trucks (as they do the vehicles of all the other emergency aid organizations). The cargos are stolen, the vehicles burned, the drivers
sometimes killed. All the men and women working for the WFP
(and for the ICRC, Action Against Hunger, Oxfam, and other
NGOs in the same field) unquestionably deserve profound respect, for they risk their lives every day.

The WFP is a forbiddingly complex organization. It manages
emergency food depots on five continents. When the prices of staple foods are low on world markets, the WFP stockpiles thousands
of tons of reserve supplies. It maintains a fleet of five thousand
trucks, with handpicked drivers.

In many countries, the WFP is compelled to resort to subcontractors, as in North Korea, for example, where the army holds
a monopoly (and hence total control) over transportation. In
other countries, only local haulers know the routes well enough—
fraught as they are with hazards, potholes, side roads—to see
that aid gets safely to its destination. This is especially true in
Afghanistan.

The WFP's Logistics and Transport Service in Rome also
maintains a fleet of aircraft. In South Sudan, hundreds of thousands of starving people live in areas inaccessible by road or river.
Cargo jets must therefore airdrop crates of food, whose safe landing is ensured by parachutes. The WFP air fleet is famous within
the UN. Many other UN departments make use of it, since it is
renowned for the reliability of its aircraft and the acrobatic skills
of its pilots. In western Sudan, for example, tens of thousands of
soldiers and police officers from African Union member states (especially Rwanda and Nigeria) struggle to maintain security for
the seventeen displaced-persons camps in the three provinces of
Darfur where war is raging. Their actions are coordinated by

the UN Department of Peacekeeping Operations in New York (DPKO). But it was WFP jets that the DPKO used to transport the African Union troops and police officers to Darfur.

In Central and South Asia, in the Caribbean, in East and Central Africa, I have witnessed the WFP's emergency response programs. I have had numerous encounters with the programs' directors and those who work under them, who are all often truly exceptional people. My admiration for the WFP is rooted in such encounters.

Daly Belgasmi is a member of a Yemenite tribe that immigrated centuries ago to central Tunisia. Born in Sidi Bouzid (the city where the recent Tunisian revolution was born on December 17, 2010), Belgasmi is a man of volcanic temper and contagious joie de vivre, endowed with remarkable determination in a fight. Trained as a nutritionist, Belgasmi has battled for nearly thirty years against the demon of hunger. In 2002, Belgasmi was the WFP's director of operations in Islamabad. Famine was raging in south and central Afghanistan. Men, women, and children were dying by the thousands. In this period, the American high command twice bombed the WFP's main food depot in Kandahar and burned it down—a warehouse that was, moreover, clearly marked with the UN flag, and whose location had been duly communicated by the WFP offices in Rome to U.S. Air Force headquarters in its underground base in Colorado. But since southern Afghanistan, and especially Kandahar, was "infested" with Taliban fighters, the American generals feared that the WFP food might fall into their enemy's hands.

As the famine in Afghanistan became ever more deadly and the food blockade imposed by the American forces ever more impenetrable, Belgasmi made a decision. He assembled a convoy of thirty 25-ton WFP trucks in Peshawar loaded with rice and wheat, crates of powdered milk, and containers of water. To the American colonel who was his usual contact at the operational headquarters in Kabul, Belgasmi sent the following message: "Our trucks will cross into Afghan territory tomorrow morning around 7:00 A.M., coming through the Khyber Pass and taking the road to Jalalabad. Please inform the air force operational command. I request, along the route whose coordinates are attached, until tomorrow evening at nightfall, a total cessation of bombing."

At dawn on the appointed day, Belgasmi gave the signal to depart. The response from the American colonel did not reach him until the convoy of trucks was already past the Torkham Gate and traveling through Afghan territory. The colonel demanded that Belgasmi immediately cancel the trip. The WFP trucks continued to descend the winding mountain switchbacks on the road to Jalalabad. Belgasmi was seated in the cabin of the first truck.

Many years later, I learned about this incident directly from Jean-Jacques Graisse himself; Graisse is the WFP's senior deputy executive director and chief of operations and its leading light. I said, "But Daly could have died!" Laughing, Graisse replied, "Even worse, if he had lost even one truck, we would have fired him right away!"

Belgasmi is currently the WFP regional director for the Middle East, Central Asia, and Eastern Europe, based in Cairo. Like a lion, he fights almost every day against the Israeli officers in Karni, the crossing point for WFP aid convoys on the border between Israel and Gaza. Every truck that makes the crossing successfully and brings aid to the undernourished men, women, and children of Gaza constitutes for Belgasmi a personal victory.

Another extraordinary person I have met at the WFP is James T. Morris, who is not at all like the stereotypical Americans Europeans know—and love. Tall, heavyset, white-haired, a friendly giant, this midwesterner was dropped into the executive director's position of the WFP by his longtime friend President George W. Bush in 2002. A billionaire, Jim Morris was a successful businessman in Indianapolis. He had served in the Indianapolis city government, had worked for several not-for-profit organizations, and was a major contributor to Bush's presidential campaign. The White House owed Morris the job of his choice. Cabinet minister? Morris said no, he wanted to travel. Ambassador? Not important enough, to his way of thinking.

How about director of a big international organization—
the WFP?

A quiet man and a loving grandfather, full of curiosity and a ferocious will to do good, Morris landed in Rome like an astronaut
landing on the moon. He knew absolutely nothing about world
hunger and the fight led by the WFP. Morris had barely assumed
his duties when he embarked on a world tour, visiting eighty countries where the WFP is active. Morris visited dozens of Food for
Assets worksites and hundreds of emergency nutrition centers
where children are treated with intravenous nutrition therapy
and, in most cases, slowly restored to life. He visited schools and
kitchens where school meals are prepared; he studied the statistics
on victims of hunger. He saw dying children, desperate mothers,
and fathers with empty eyes.

And he was sorely troubled. I remember one expression he used
over and over: "This cannot be. . . ."

Drawing on his formidable energy and his vast experience as
a businessman who had built an empire, Morris threw himself
into his work. Morris is a Christian, a member of the Episcopal
Church. Right in the middle of his stories, I sometimes saw tears
in his eyes. Rereading some of the letters that he sent me, I find
words that perfectly sum up what moves him:

> Dear Mr. Ziegler,
>   Thank you for all the good that you do. I appreciate your
> efforts on behalf of the world's poor and hungry. . . . So
> many people have need of us, it is so sad, above all for the
> little ones.
>   Good luck,
>   Jim

Or this, from another letter:

> Every one of us must do all that we can for others, every day,
> whether they are near or far away from us. All I know is, the
> thing that unites us is our humanity. . . . It is impossible to

understand the great mystery of life. . . . There is so much to
be done, so few of our efforts succeed.

A friendly relationship, with rather comic political consequences, grew up between Morris and me. It was Graisse who introduced us over lunch in the Port Gitana restaurant in Bellevue,
on the lakeshore just north of Geneva. Morris had invited me as
a special guest to the quadrennial WFP conference in June 2004
in Dublin. Every four years, in fact, the WFP gathers together
all its regional directors to discuss proposed strategies for the
organization.

The Josué de Castro era had ended decades ago, and no one at
the WFP (or at the FAO) remembered the idea of a right to food
any longer. Within the UN system, human rights had become the
province of the Human Rights Council, not of the UN's specialized agencies. The WFP considered itself to be a humanitarian
aid organization, period.

In Dublin, I pleaded on behalf of a normative, rights-based
approach, and accordingly for structural economic and social
changes. Belgasmi, Graisse, and Morris supported me. On June
10, the last day of the conference, Morris put to a vote a resolution "on the rights-based approach to hunger" stipulating that
henceforth the realization of the right to food would constitute the
strategic goal of the WFP.

Throughout this period, as I have explained, whenever I presented my semiannual reports and formulated my recommendations to the Human Rights Council in Geneva and to the Social,
Humanitarian and Cultural Affairs Committee (also known as
the Third Committee) of the UN General Assembly in New York,
the various American ambassadors would attack me virulently.
They denied the very existence of any human right to food.

Summoning all his resources of energy and diplomatic skill,
Morris, on the other hand, would from now on defend this right.
And yet, as executive director of the WFP, Morris was regularly
invited to appear before the UN Security Council to report on
the world food situation. During his presentations, Morris twice

quoted me, referring to me as "my friend Jean Ziegler, whose political opinions I do not share."

This situation in fact particularly perturbed Ambassador Warren W. Tichenor, George W. Bush's special envoy in Geneva. Soon he no longer dared to come to meetings of the Human Rights Council. He sent instead his adjunct, the sinister Mark Storella, who, of course, continued to attack me. In the eyes of the American mission's diplomats in Geneva, as in the eyes of their colleagues in New York, I remained a crypto-communist abusing his UN mandate whom they claimed to have unmasked: "You have a secret plan!" "You are engaged in a secret crusade against our president's policies!" How many times did I hear these idiotic accusations?

They demanded my dismissal numerous times. But the friendship of UN secretary general Kofi Annan and the diplomatic savoir faire of UN High Commissioner for Human Rights Sérgio Vieira de Mello ultimately saved my mandate. The last time, however, just barely so . . .

For Ambassador Tichenor, Morris was beyond reach. A Republican Party heavyweight, a businessman free of any ties to the administration, Morris could at any moment pick up his phone to call the White House. I don't know if he ever spoke about the right to food with his friend George W. Bush.

Exhausted, worn out, Jim Morris left Rome in the spring of 2007.

18

VICTORY OF THE PREDATORS

In all my years as Special Rapporteur on the Right to Food, the finest moments—the most intense and the most moving—were those I spent in school cafeterias and kitchens in Ethiopia, Bangladesh, Mongolia, and many other countries. There, I felt proud to be human.

The food varied according to the country. Meals were prepared with local products: manioc, teff, and millet in Africa; rice and chicken with sauces in Asia; quinoa and sweet potatoes on the high Andean plateau. On every continent, the WFP meals included vegetables. For dessert, there was always local fruit: mangoes, dates, grapes, and so on, depending upon the country.

One daily meal served in the school cafeteria could prompt parents to send their children to school and to keep them there. School meals obviously promoted learning and enabled students to concentrate on their studies. For only about 25 cents, the WFP could fill a bowl with porridge, rice, or vegetables, and give students a monthly ration to take back home with them. Fifty dollars was enough to feed a child in school for a year. In most cases, children were given breakfast and/or lunch at school. These meals were prepared at the school itself, by the community, or in central kitchens. Some school cafeteria programs offers full meals, while others provided high-energy biscuits or healthy snacks. The

WFP's famous take-home rations completed the cafeteria programs. Thanks to this system, entire families received food when their children went to school. Food was purchased in the area as much as possible; this approach profited local small farmers. Moreover, the meals served in the cafeteria were fortified with micronutrients. In this way, by providing essential nutrition in the poorest regions, school meals sometimes succeeded in breaking the cycle of hunger, poverty, and the exploitation of children. School meals were also given to children living with HIV/AIDS, orphans, handicapped children, and demobilized child soldiers.

Before 2009, the WFP provided school meals to 22 million children on average, the majority of them girls, in some seventy countries and at a total cost of $460 million. In 2008, the WFP provided take-home rations to 2.7 million girls and 1.6 million boys. The WFP fed 730,000 children in kindergartens in fourteen countries: Haiti, the Central African Republic, Guinea, Guinea-Bissau, Sierra Leone, Senegal, Benin, Liberia, Ghana, Kenya, Mozambique, Pakistan, Tajikistan, and the Occupied Palestinian Territories.

One day, in a school in Jessore, in Bangladesh, I noticed, all the way in the back of the class, a boy about seven years old who was sitting with his plate of porridge and beans in front of him on his desk, without touching it. He sat motionless, his head lowered. I asked Shah Mushid, who was then head of the WFP Jessore sub-office (he is now head of human resources and training for WFP Bangladesh), about the boy. He replied evasively, obviously embarrassed. Finally he allowed, "There are always problems. . . . Here in Jessore we don't have the means to give the students family rations that they can take home. So the boy refuses to eat. . . . He wants to take his meal home to his family."

I was shocked. "But why don't you let him? It's because he loves his family!"

Murshid replied, "The boy is hungry. He has to eat. The rules don't allow taking food out of the school."

This problem recurs wherever the WFP sets up school

cafeterias. Where the WFP's budget (and the funds from the NGOs who support it) does not allow it to provide supplementary food for students to take home to their families, strict rules apply.

In Sidamo, in southern Ethiopia, for example, the teacher locks the cafeteria as soon as a meal has been served, in order to force the students to eat on-site. When the children leave the cafeteria and head for the row of water taps in the courtyard to brush their teeth and wash their hands, the teacher reenters the school to check that all the meals have been eaten and that there are no full or half-full plates hidden under the children's desks.

The children love their families. To eat while their loved ones back at home are hungry conflicts with their feelings of loyalty and solidarity. So some of them prefer going without, gnawed by hunger, rather than eating, gnawed by guilt.

However, for tragic reasons, this problem hardly exists anymore. What happened is this: one afternoon in early October 2008, the seventeen leaders of the Eurozone governments met in the Élysée Palace in Paris. At six o'clock, Angela Merkel and Nicolas Sarkozy appeared on the steps before the press. To the assembled journalists they declared, "We have just freed up 1.7 trillion dollars to unfreeze the interbank lending market and to increase the bank's minimum capital requirements from 3 to 5 percent." Before the end of 2008, subsidies from the Eurozone countries for emergency food aid fell by almost half.

The WFP budget prior to the financial crisis was usually around $6 billion; in 2009 it fell to $3.2 billion. The WFP had to practically suspend school meals worldwide, particularly in Bangladesh. More than a million little Bengali girls and boys have since been deprived of WFP school meals.

In 2005, I visited many schools in Dhaka, Chittagong, and elsewhere (I discuss my 2005 mission to Bangladesh further in chapter 20, "Jalil Jilani and Her Children"). It was obvious that the little kids I saw, with their big black eyes and their frail bodies, were getting their only consistent meal of the day at school. I remember too a meeting I had that lasted several hours in the office of the minister of education in Dhaka. My colleagues and I, supported by the local representative of the UN Development Programme, were fighting tooth and nail to end the practice of closing Bengali schools during their long vacations: in other words, to ensure the children access to one daily meal twelve months a year. The minister refused. Today, the question is a dead issue because, in one country after another, the WFP has suspended its school meals programs.

For 2011, the WFP estimated its "incompressible needs" (minimum budgetary requirements) at $7 billion; as of early December 2010, it had received $3.7 billion. This shortfall in revenue has had tragic consequences.

I saw the results close up in Bangladesh. In 2009, in this poor, densely populated country prone to natural disasters, 8 million men, women, and children lost all their sources of income and were therefore, to use the WFP's own term, "on the edge of starvation," owing to two back-to-back catastrophes: the devastation of agriculture caused by an extremely violent monsoon, and the closure of many textile factories reeling under the full brunt of the global financial crisis. In that year, the WFP's Asian office requested $257 million to provide aid to Bangladesh. It received $76 million. The situation was even worse in 2010: the Asian office received only $60 million for Bangladesh. For 2011, it expected an even more serious further collapse in donor states' subsidies— and thus an even greater number of people condemned to die of hunger.

In other parts of the world, the situation is equally tragic. In 2010, the WFP budget for sub-Saharan Africa was $2.6 billion— $1.1 billion less than the agency needed to accomplish its mission.

It would be obviously unfair in any way to reproach Merkel, Sarkozy, José Zapatero (the prime minister of Spain from 2004 to 2011), or any of the other government leaders associated with the decision taken in 2008 to pour 1.7 trillion euros into their banks, to the detriment of subsidies allocated to the WFP. Merkel and Sarkozy were elected to support, and if necessary to restore order to, the German and French economies. They were not elected to fight world hunger. However, the suffering children of Chittagong,

Ulan Bator, or Tegucigalpa don't vote. Nor will they die on the Avenue des Champs-Élysées in Paris, on the Kurfürstendamm in Berlin, or in the Plaza de Armas in Madrid.

The ones who are truly responsible for this situation are the speculators—hedge fund managers, bigwig bankers, and other predators of the globalized finance industry who, by their obsession with profit and personal gain, wrecked the global financial system and destroyed billions and billions of dollars' worth of countries' national wealth. These predators should be tried for crimes against humanity. But they are so powerful—and governments are so weak—that they are obviously under no threat of anything like that at all.

On the contrary. Since 2009, they have merrily returned to their wicked ways, barely checked by the few timid new laws and standards—minimum capital requirements, lightly enhanced oversight of derivatives, and so on—announced by the Basel Committee on Banking Supervision, the institution that coordinates the rich countries' central banks. Almost as if nothing had ever happened.

19

"NATURAL" SELECTION REDUX

In the WFP's dilapidated building in Rome there are two rooms where the fate—or, more concretely, the life or death—of hundreds of thousands of people is decided every day.

The first of these, the situation room, houses the WFP's database. The WFP's greatest power resides in its capacity to react as quickly as possible to disasters and to mobilize with minimal delay the ships, trucks, and planes needed to transport food and water indispensable to the survival of the victims of hunger. The WFP's average reaction time is about forty-eight hours. The walls of the situation room are covered with enormous maps and screens. On the long black tables are piles of meteorological charts, satellite images, and so on. All the harvests everywhere in the world are monitored on a daily basis. The movements of locust swarms, the tariffs on maritime freight, and the prices of rice, wheat, corn, millet, barley, and palm oil on the Chicago Board of Trade commodities exchange and other agricultural commodities exchanges around the world, as well as many other economic variables—all are constantly scrutinized, studied, and analyzed. En route from Vietnam and the port of Dakar, for example, rice is at sea for six months. Changes in the cost of transportation play a crucial role. The predictable variations in the price of a barrel of oil constitute another element that is closely followed by the economists and specialists in transportation insurance who work in the WFP situation room. These specialists are highly effective, ready to deliver any information necessary even with the least advance warning.

The other strategy room at the WFP headquarters in Rome, even if much less impressive at first sight, and less busy with experts of all kinds, is the Vulnerability Analysis and Mapping (VAM) branch of the Food Security Analysis Unit, currently headed by the energetic Joyce Luma. There Luma's team issues minutely detailed analyses that identify vulnerable groups on all five continents.

In a certain sense, Luma is tasked with establishing the hierarchy of extreme poverty. She works with all the other UN organizations, NGOs, churches, national ministries of health and of social affairs, and above all the regional and local WFP directors. In Cambodia, Peru, Bangladesh, Malawi, Chad, Sri Lanka, Nicaragua, Pakistan, Laos, and many other countries, she subcontracts field research to local NGOs. Armed with detailed questionnaires, the field researchers (usually women) go from village to village, shantytown to shantytown, hamlet to hamlet, interviewing heads of families, isolated individuals, and single mothers about their income, employment, food situation, illnesses afflicting their family, lack of water, and so on. The questionnaires generally comprise between thirty and fifty questions, all developed in Rome. Once they have been filled out, the questionnaires are returned to Rome to be analyzed by Luma and her team.

Elie Wiesel is certainly one of the greatest writers of our time. He is himself a survivor of the camps at Aushwitz-Birkenau and Buchenwald. He has highlighted with particular clarity the nearly insurmountable contradiction that affects any discussion of the extermination camps. On the one hand, the Nazi camps represent a crime so monstrous that no human speech is really capable of expressing it: to speak of Auschwitz is to make the inexpressible banal. But on the other hand there is the unavoidable obligation of memory: everything, even the most monstrous crime, may happen again at any time. It is therefore necessary to speak out, informing and warning the generations who have not experienced the unspeakable of the threat of relapse.

At the heart of the Nazi horror was the "selection process." The ramp at which prisoners disembarked from the trains at Auschwitz was the place where, in the blink of an eye, the fate of each new arrival was decided: to the left for those who would die, to the right for those who, for a while at least, would be allowed to live.

Selection is equally at the heart of Luma's work. Since the WFP's funding has collapsed and the amount of available food going forward will be insufficient to respond to the needs of the millions whose empty hands reach out for it, one has no choice but to choose.

Luma tries to be fair. By every technical means at the disposal of the biggest humanitarian organization in the world, she strives to identify, in each country ravaged by hunger, the most gravely afflicted people, the most vulnerable, those in the most immediate danger of starving to death. The individuals and groups who, unluckily, do not fall into the "extremely vulnerable" category are left stranded, although they belong no less among the populations threatened with serious undernutrition—and thus with impending death, however much delayed.

Joyce Luma, this woman radiant with humanity and compassion, decides who will live and who will die. She too practices "selection," even if she does so—and even if this fact forbids any comparison with the Nazi horror—in the name of an objective necessity imposed upon the WFP.

20

JALIL JILANI AND HER CHILDREN

Bangladesh is an immense fertile delta some 144,000 square kilometers (55,600 square miles) in area, with a population of 150.5 million. It is the most densely populated country on the planet. Before my first mission to Bangladesh, Dr. Ali Toufiq Ali, the Bengali permanent representative to the UN office in Geneva, told me, "You will never be alone, you will see people everywhere." And in fact, wherever I went, from north to south, in Jessore or Jamalpur, or in the mangrove swamps on the Bay of Bengal, I found myself surrounded by smiling men, women, and children, wearing clothes that, though worn, were impeccably clean and pressed. Unassuming, smiling, constantly in motion, the Bengalis are indeed *everywhere*.

But Bangladesh is also one of the most corrupt countries in the world. Throughout my entire mandate as UN Special Rapporteur, I was only once offered a bribe, in Dhaka, in point of fact, in 2005. Accompanied by Christophe Golay and my colleagues Sally-Anne Way and Dutima Bagwali, two brilliant and elegant young women, I was seated before the minister of foreign affairs—a fat man with cruel eyes, covered in sweat despite the ceiling fan, and himself one of the country's textile barons—and his parliamentary secretary, in the ministry's main reception hall. For at least an hour, I had been trying to get the minister to talk about the vast shrimp farms that Indian multinational corporations had been authorized to develop in the mangrove swamps along the coast of the Bay of Bengal.

Bengali fishers had complained to me. Their traditional small-scale coastal fisheries had been ruined, they told me. The Indian shrimp farms were blocking their access to the sea along hundreds of miles of coastline. I was confronted with an obvious violation of the Bengali fishers' right to food on the part of their own government. I wanted to obtain from the minister a copy of the contracts signed by his government and the Indian multinationals involved.

The minister totally stonewalled me. Instead of replying to my questions, he insisted on embarking—very clumsily—on a charm offensive directed toward my pretty young colleagues, which, very obviously, exasperated both of them.

Suddenly the minister smiled sweetly and, in front of his parliamentary secretary, said, "My company periodically offers high-level conferences to its international clients. I invite intellectuals and university professors from all over the world, mostly from the United States and Europe. Our clients appreciate it. Those who attend our conferences do too. We pay sizeable honoraria. . . . Do you have any free time in your calendar? I would be happy to invite you."

A young Guyanese woman with a fiery temper, Bagwali had already stood up. Way and Golay were likewise ready to head out the door. I restrained them. The parliamentary secretary smiled devotedly. The minister did not understand why, in such an abrupt fashion, I put an end to our meeting and we took our leave.

Dhaka . . . The wet heat makes your clothes stick to your skin. At the Ministry for Economic Cooperation and Development I met with Waliur Rahman, then the secretary of the Ministry of Foreign Affairs, As a young student, he had been sent by Mujibur Rahman to Geneva in 1971, during the Bangladesh Liberation War, when the country (then East Pakistan), with help from India, fought off an occupying Pakistani army and seceded from Pakistan.

Muammar Murshid and Rane Saravanamuttu, from the local WFP bureau, joined Waliur and me for a visit to the Gulshan shantytown, part of the Karall slum, one of the largest in Dhaka, where 800,000 people live in shacks and huts made from canvas and planks on the muddy riverbanks. All the peoples of this vast "land of a thousand rivers," as the Bangladeshis call their splendid homeland, are gathered here: thousands of refugee families from Jamalpur, where the monsoon had caused twelve thousand deaths the year before; Shantali and other tribal peoples of the mangrove forest regions; and members of animist tribal groups, the poorest inhabitants in the country and the most despised by the Muslim majority. In the Gulshan shantytown there also live hundreds of thousands of members of the urban underclass, the permanently unemployed, and workers recently laid off from the gigantic textile subcontractors. Adherents of all the nation's religions mingle together here too: Muslims, the vast majority; Hindus from the north; and Catholics, members of formerly animist tribal groups who were converted by European missionaries during the colonial period.

I asked to visit some of the people's shacks. Rahman contacted the municipal ward commissioner representing the neighborhood. Few of the dwellings in the shantytown have doors. A simple colored curtain covered the entrance. The commissioner lifted the curtain. In the room, feebly lit by a candle, I found, sitting on the single bed, a young woman wearing an old sari with four small children. They were thin and pale. They stared at us with big black eyes, neither speaking nor moving. Only on the young mother's face was there the hint of a timid smile. Her name was Jalil Jilani. Her children were two, four, five, and six years old. Two girls, two boys. Her husband, a rickshaw driver, had died of tuberculosis a few months before.

Bangladesh is one of the main countries in South and Southeast Asia where Western multinational textile corporations have their jeans, sport shirts, suits, and other clothing sewn, mainly by women, in so-called free or special economic zones. The costs of production are unbeatable. The factories that subcontract the work are mostly owned by South Korean and Taiwanese businesses.

The free economic zones comprise almost all the suburbs south of Dhaka, where immense concrete buildings rising from seven to

ten stories jostle against one another. No health and safety regula-
tions, no minimum wage laws are enforced. Unions are banned.
Workers are hired and fired according to the fluctuations in orders coming from New York, London, Hong Kong, or Paris.

Jilani had been employed by Spectrum Sweater Industries, in
Savar, near Dhaka. More than five thousand people, 90 percent of
them women, cut, sewed, and packaged T-shirts, sweatpants, and
jeans at Spectrum for big American, European, and Australian de-
signer labels. The legal minimum wage in urban areas in Bangladesh
in 2005 was 930 takas a month, or about $14.50. Spectrum Sweater
paid its workers 700 takas per month, or about $11.00.

The Clean Clothes Campaign (CCC), the global garment industry's largest alliance of labor unions and NGOs, founded in 1989
and based in Amsterdam, is "dedicated to improving working conditions and supporting the empowerment of workers in the global
garment and sportswear industries." The CCC has calculated that
of the $75 price of a pair of jeans made at Spectrum Sweater, the
garment worker who sewed them was paid about 33 cents.

On the night of April 10 to 11, 2005, the nine-story reinforced-
concrete building housing Spectrum Sweaters collapsed. The
cause: flawed construction and a lack of maintenance and in-
spections. But in the free economic zones, the factories operate
twenty-four hours a day. As a result, when the disaster struck, every workstation was occupied. When the building fell, hundreds
of workers went down with it and were buried alive beneath the
rubble. The government refused to provide an exact figure for the
number of victims. Spectrum Sweater, for its part, laid off all the
survivors, without paying any compensation or severance.

The extreme undernutrition Jilani and her children were suffering was evident at first sight. I turned to Muammar Murshid.
He shook his head. No, the young mother and her children were
not on the list of recipients of WFP food aid. The reason? Jilani
had been laid off in April. Murshid was very sorry. He was the
WFP representative in Bangladesh. He had to enforce the organization's directives from Rome. Jilani had had regular work during
more than three months of the current year, which automatically

excluded her from WFP aid. She had no grounds for appeal. In the accounting system for extreme poverty that Joyce Luma oversees in Rome, Jalil Jilani and her four children, gnawed by hunger, had thus exited the category of those with a right to aid.

Murshid murmured a quick good-bye in Bengali. I left all the takas I had on me at the end of the bed. Rahman let the curtain fall behind us.

21

THE DEFEAT OF JACQUES DIOUF

The FAO is sumptuously housed. Surrounded by fragrant gardens and parasol pines, its palatial world headquarters, a vast modernist building on the Viale delle Terme di Caracalla, formerly housed Mussolini's colonial ministry, the Department of Italian East Africa. Until recently, a great treasure graced the square in front of the building: the Obelisk of Axum, which was returned to Ethiopia in 2005.

Founded, as I have said, at the instigation of Josué de Castro and his friends in October 1945 (that is, one and a half years after the UN), the FAO was given an ambitious mandate, outlined in the first article of organization's constitution:

> *Article I*
>
> *Functions of the Organization*
>
> 1. The Organization shall collect, analyse, interpret and disseminate information relating to nutrition, food and agriculture. In this Constitution, the term "agriculture" and its derivatives include fisheries, marine products, forestry and primary forestry products.
>
> 2. The Organization shall promote and, where appropriate, shall recommend national and international action with respect to:

(a) scientific, technological, social and economic research relating to nutrition, food and agriculture;

(b) the improvement of education and administration relating to nutrition, food and agriculture, and the spread of public knowledge of nutritional and agricultural science and practice;

(c) the conservation of natural resources and the adoption of improved methods of agricultural production;

(d) the improvement of the processing, marketing and distribution of food and agricultural products;

(e) the adoption of policies for the provision of adequate agricultural credit, national and international;

(f) the adoption of international policies with respect to agricultural commodity arrangements.

In the vast white-marble-clad entrance hall of the FAO's headquarters, the agency's insignia is mounted on the right-hand wall. Beneath a stalk of wheat on a blue ground is written the organization's motto: *Ecce panis*—"Let there be bread" ("for all" is understood).

What, today, is the situation of the FAO?

The organization comprises 191 member states (as well as the European Union, the Faroe Islands, and Tokelau as associate members). However, global agricultural policy, and in particular the question of food security, is determined by the World Bank, the IMF, and the WTO. The FAO is largely absent from the battlefield. It has been bled dry, gutted.

The FAO is an intergovernmental organization. The global private corporations that control most of the world market in food products fight against it. These corporations enjoy a certain influence over the policies of the principal Western governments. The result: these governments withdraw from the FAO, restricting its budget and boycotting the world conferences on food security that the FAO convenes in Rome.

Currently about 70 percent of the FAO's meager funds serve to pay its staff. Of the remaining 30 percent, about half goes to pay the fees of the FAO's myriad external "consultants." Only the remaining 15 percent of the FAO's budget pays for technical cooperation, agricultural development in the South, and the fight against hunger.

For several years, the organization has been the object of virulent criticism, largely unjustified, since it is the governments of the industrialized countries that deprive the FAO of its capacity for taking action. In 1989, the English writer Graham Hancock published a book, which has since been reprinted many times, titled *The Lords of Poverty*. According to Hancock, the FAO is nothing but a grim, gigantic bureaucracy that, on account of an interminable succession of conferences, meetings, committees, and expensive events of all kinds, does nothing but administrate poverty, undernutrition, and hunger. In its day-to-day practice, the bureaucracy at the Baths of Caracalla, Hancock says, incarnates the very opposite of the project initially conceived by de Castro. Hancock's conclusion is unanswerable: "One gets the sense . . . of an institution that has lost its way, departed from its purely humanitarian and developmental mandate, become confused about its place in the world—about exactly what it is doing, and why." *The Ecologist* is even more scathing. In a special issue published in 1991, the magazine assembled a collection of essays written by respected international experts such as Vandana Shiva, Edward Goldsmith, Helena Norberg-Hodge, Barbara Dinham, and Miguel A. Altiera under the title *The UN Food and Agriculture Organization: Promoting World Hunger*. The authors accuse the FAO of mistaken strategies, wasting colossal sums of money that are swallowed up by useless action plans, and a range of false economic analyses whose effect has been not to reduce but to increase the tragedy of world hunger. As for the BBC, its verdict on the periodic summits organized by the FAO is equally unanswerable: the summits are simply a "waste of time"—and money.

In my opinion, even if the FAO must admit that certain of these criticisms are valid, the organization must be defended against any and all of its critics—and especially against the giants of the agri-food trade, which have their tentacles everywhere, and their accomplices in the Western governments.

In 2010, the industrialized countries represented in the

Organisation for Economic Co-operation and Development (OECD) spent $349 billion on subsidies for agricultural production and export. Export subsidies in particular are responsible for the dumping of agricultural products practiced by the rich countries in the markets of the poor countries. In the southern hemisphere, such dumping creates severe poverty and hunger.

By comparison, the FAO's "total proposed *Net Budgetary Appropriation* to be funded from assessed contributions" for the 2012–13 fiscal year is $1.0571 billion—a fraction of the amount that the rich countries spend on their agricultural subsidies.

How can the organization meet its obligations, at least partially? The term *monitoring* is used by the FAO to designate a strategy of transparency, information sharing, and permanent investigation into the details of the evolving global situation in undernutrition and hunger. On all five continents, vulnerable groups are listed and ranked, month by month; the various deficiencies in micronutrients (vitamins, minerals, trace elements) are logged, element by element, region by region. Statistics, graphs, reports flow uninterruptedly from the FAO's headquarters in Rome: not one of the immense army of the hungry suffers or dies without leaving a trace on an FAO graph.

The self-proclaimed adversaries of the FAO also criticize its monitoring policies. Instead of compiling minute statistics on the hungry, they say, constructing mathematical models of suffering, and drawing colored graphs to represent the dead, the FAO would do better to use its money, its know-how, and its energy to reduce the number of victims. This critique also seems to me unfair. The FAO's monitoring informs our anticipatory consciousness; it prepares the way for a future uprising in our collective conscience. For one thing, this book could not have been written without the statistics, inventories, graphs, and other tables produced by the FAO.

The FAO owes its monitoring system to one man in particular, Jacques Diouf, of Senegal, the organization's director general from 2000 to 2011. Diouf is a nutritionist, and a socialist who served as secretary of state for science and technology from 1978 to 1983 under both Léopold Sédar Senghor and his successor, Abdou

Diouf. Previously, Diouf had served as the first executive secretary of the West Africa Rice Development Association (WARDA; now the Africa Rice Center), based in Monrovia, Liberia.

Good-humored, gifted with a subtle intelligence and formidable vitality, Diouf woke up and shook up the FAO bureaucrats in Rome. His aggressive, sometimes brutal way of addressing heads of state, and his statements, issued through newspapers and on radio and television worldwide, that attempted to arouse and influence public opinion in the dominant countries deeply irritated certain leaders of Western governments and other officials. Many sought any possible pretext to discredit him.

One such attempt to discredit Diouf took place at the second World Food Summit, held by the FAO in Rome in 2002. On the top floor of the FAO headquarters building, the director general has access to a private dining room where, like all the directors of the UN's specialized agencies, he entertains heads of state and heads of government. On the third day of the summit, the day after an especially harsh speech by Diouf criticizing the private global agri-food corporations, the British press published front-page stories featuring details of the menu from the dinner Diouf had hosted the day before for the visiting heads of state and heads of government. The meal, obviously, had been lavish. The head of the British delegation, who had himself attended the dinner, took this "revelation" as a pretext for launching an incendiary diatribe against the director general in an open plenary meeting, accusing him of speaking about hunger in public while, in private, stuffing himself at the expense of the FAO's contributing member nations.

I feel admiration for Jacques Diouf, because I have seen him at work on many occasions. For example, in July 2008, following the first of a series of surges in the prices of staple food commodities on the global market, food riots raged in thirty-seven countries, as I have said above. The regular session of the UN General Assembly was set to open in September. Diouf was convinced that it was necessary to seize this occasion to launch a massive international campaign whose aim would be to paralyze the activity of the speculators who were driving up food prices. He therefore mobilized his friends in the Socialist International, a worldwide organization of social democratic, socialist, and labor parties. The Spanish government under Zapatero agreed to spearhead this campaign: the resolution that would be proposed on the first day of the General Assembly session would be sponsored by Spain.

Foreseeing the battle to come, Diouf furthermore convened a meeting of the leaders of all the international organizations involved in the fight against hunger and related to one of the more than one hundred member parties of the Socialist International. The meeting took place at the seat of the Spanish federal government, the Palacio de la Moncada, in Madrid. In the great white room, lit by the Castilian sun, there were, seated around a black table, António Guterres, former president of the Socialist International, former prime minister of Portugal, and then UN High Commissioner for Refugees; the French socialist Pascal Lamy, director general of the WTO; leaders of the Brazilian Workers' Party; a cabinet minister in the British Labour government; obviously, José Zapatero himself, together with Miguel Ángel Moratinos, his minister of foreign affairs, and Bernardino León, the Spanish secretary of state for foreign affairs; and finally myself, in my capacity as vice president of the UN Human Rights Council Advisory Committee.

Diouf shook us like a hurricane. Linking a whole series of precise measures against speculators to a demand addressed to the states that are signatories to the Covenant on Economic, Social and Cultural Rights, reminding them of their obligation to honor the right to food, Diouf's proposed resolution incited intense discussions among everyone present. Diouf hung on. Agreement was reached toward two o'clock in the morning.

In September, before the UN General Assembly in New York, Spain presented its resolution, seconded by Brazil and France. But the measure was simply swept aside by a coalition led by the representative from the United States and several ambassadors under the remote control of certain global agri-food corporations.

POSTSCRIPT: THE MURDER OF IRAQ'S CHILDREN

Obviously, neither the WFP nor the FAO can be held responsible for the difficulties and the setbacks that they have encountered.

But there is one case in which the UN itself, willingly and intentionally, has caused the extermination by hunger of hundreds of thousands of human beings. This crime was committed within the framework of the oil-for-food program imposed upon the Iraqi people by the UN Security Council in 1995 and ostensibly discontinued with the U.S. invasion of Iraq in 2003 (the program continued de facto until 2010).

Recall the history behind the program. On August 2, 1990, Saddam Hussein sent his forces to invade the emirate of Kuwait, which he annexed, proclaiming it the twenty-seventh Iraqi province. The UN initially responded by decreeing an economic blockade against Iraq and by demanding the immediate withdrawal of Iraqi forces from Kuwait; then the UN issued an ultimatum set to expire on January 15, 1991.

Under the direction of the United States, Western and Arab countries formed a coalition whose forces attacked the occupying Iraqi troops in Kuwait once the UN ultimatum had expired; 120,000 Iraqi troops and 25,000 Iraqi civilians lost their lives in Kuwait. But the tanks under commander in chief General Norman Schwarzkopf halted a hundred kilometers from Baghdad, leaving intact the Republican Guard, the Iraqi dictator's elite troops. The fall of Saddam Hussein was considered likely to lead to the installation in Baghdad of a majority Shi'ite government; the Western governments feared the Iraqi Shi'ites like the plague, suspecting them of owing allegiance to the tyrannical regime in Teheran.

The UN intensified its blockade, but at the same time inaugurated the oil-for-food program (officially, the Office of the Iraq Programme—Oil-for-Food), which permitted Saddam Hussein to sell a certain quantity of Iraq's oil on the world market every six months. (With 112 billion barrels of oil, Iraq has the second-largest petroleum reserves in the world, after Saudi Arabia, with 220 billion barrels, and ahead of Iran, with 80 billion barrels.) The revenues from the sales were paid into an escrow account held until 2001 by BNP Paribas in New York. The money allowed Iraq to buy on the world market goods indispensable to the survival of its population. In concrete terms, a business holding a contract for delivery of goods to the Iraqi government submitted a request for the liberation of the necessary funds to the bank in New York. The UN approved or denied the delivery according to the criterion of suspected "dual-use function": if the UN believed that any good—machines, replacement parts, chemicals, construction materials, and so on—might have any military use, the request was denied.

The coordinator of the oil-for-food program, who headed the UN Office of the Humanitarian Coordinator in Iraq (UNOHCI), was installed in Baghdad, with the rank of UN assistant secretary general, and with eight hundred UN functionaries at his disposal as well as twelve hundred locally hired officials. He reported to the Office of the Iraq Programme (OIP) in New York, which was responsible for examining the requests presented by businesses seeking to make exports to Iraq. The program was headed by Benon Sevan of Cyprus, the former head of the UN security services, who was promoted to the rank of assistant secretary general in the UN Department of Political Affairs in July 1992 as a result of pressure from the United States and who would prove to be a crook. Sevan was in fact indicted for fraud in New York District Court but fled to Cyprus, where he is living happily ever after. The OIP was in turn overseen by a committee on sanctions of the UN Security Council responsible for the program's general strategy.

On paper, the oil-for-food program was inspired by the usual principles of embargoes as they are applied by the UN. But in practice, the program was deliberately diverted from its purpose in a way that proved deadly for the civilian population. Very soon, in effect, the sanctions committee began to deny more and more frequently requests for the importation of food, medicines, and other vital necessities under the pretexts that the food might be used to feed Saddam Hussein's army, that the medicines might contain chemicals with military applications, that certain parts of medical devices could also be used to make arms, and so on.

In the hospitals of Iraq, patients began to die for lack of medicines, surgical instruments, and sterilization supplies. According to reliable estimates, as many as 550,000 Iraqi children may have died of undernutrition and other causes between 1996 and 2000. Thus, gradually, beginning in 1996, the oil-for-food program was diverted from its original mission and came to serve as a weapon of collective punishment for the Iraqi population, based on the deprivation of food and medicine. One of the most respected international jurists, Marc Bossuyt, former chairman of the UN Sub-Commission on the Promotion and Protection of Human Rights (today the UN Human Rights Council), has said that the strategy of the UN sanctions committee against Iraq amounted to "genocide."

Here are a few statistics on the consequences of this murderous strategy imposed upon Iraq, a large country with 26 million inhabitants. According to estimates by the German NGO Medico International, less than 60 percent of medicines essential to the treatment of cancer was admitted. The importation of dialysis machines for the treatment of patients with kidney failure was purely and simply forbidden. Gulam Rabani Popal, the WHO representative in Baghdad, asked permission in 2000 to import thirty-one dialysis machines, which Iraqi hospitals urgently needed. The eleven machines that were finally authorized by the OIP in New York were held at the Jordanian border for two years.

In 1999, Carol Bellamy, the American executive director of UNICEF from 1995 to 2005, addressed the Security Council in person. The sanctions committee had refused to authorize the importation of supplies necessary for intravenous nutrition for seriously undernourished infants and young children. Bellamy protested vigorously. The sanctions committee remained adamant in its refusal.

War had destroyed the gigantic water purification plants on the Tigris, Euphrates, and Shatt al-Arab rivers. The sanctions committee refused to allow delivery of construction materials and replacement parts necessary for the reconstruction and restoration of the plants. The incidence of infectious diseases caused by the pollution of drinking water exploded.

In Iraq, summer temperatures can reach 45 degrees Celsius (113 degrees Fahrenheit). The sanctions forbid the importation of replacement parts for refrigerators and air-conditioning units. In the butcher shops, meat began to rot. Grocers watched as their milk, fruit, and vegetables were destroyed by the heat. In the hospitals, it became impossible to keep the small amount of available medicines properly refrigerated.

Even the importation of ambulances was blocked by the sanctions committee. The reason? Because they contained "communications systems that could be used by Saddam Hussein's troops." When the ambassadors of first France and then Germany remarked that communications systems such as a telephone were indispensable in every ambulance in the world, the American ambassador couldn't have cared less: no ambulances for Iraq. (A few were later permitted after a delay of a year or two.)

Tens of thousands of Egyptian *fellahin*, specialists in irrigation who bear witness to a magnificent ancestral heritage of agricultural experience acquired in the Nile River Valley and the Nile Delta, were working in Iraq in the ancient Fertile Crescent, between the Tigris and the Euphrates Rivers in Iraq. Nonetheless, Iraq in this period imported nearly 80 percent of its food. But under the embargo, food imports were usually intentionally delayed by the sanctions committee. Documents reveal that thousands of tons of rice, fruit, and vegetables went bad in trucks held up at the borders because they had not gotten the green light from New York, or because they got it only after months of delay.

The dictatorship of the sanctions committee was merciless. It attacked even the educational system. The Security Council accordingly forbade the importation of pencils. The reason? Pencils contain graphite, a material with potential military uses.

The UN blockade completely destroyed the Iraqi economy. As a report released in March 1999 under the auspices of Celso Amorim, who represented Brazil on the UN Security Council and presided over its Iraq Panels in 1998 and 1999, said in a widely quoted passage, "Even if not all suffering in Iraq can be imputed to external factors, especially sanctions, the Iraqi people would not be undergoing such deprivations in the absence of the prolonged measures imposed by the Security Council and the effects of war." Hasmy Agam, the head of the Malaysian mission to the UN, wrote in even starker terms: "How ironic it is that the same policy that is supposed to disarm Iraq of its weapons of mass destruction has itself become a weapon of mass destruction."

What can explain this drift in UN policy?

Elected in 1993, President Bill Clinton did not want under any circumstances to get involved in a second Gulf war. Under these conditions, the Iraqi people had to be subjected to a regime of such intense suffering that they would rebel against the tyrant ruling over them and drive him from power. Clinton's secretary of state, Madeleine Albright, must without doubt be held mainly responsible for the secret transformation of the oil-for-food program into a weapon of collective punishment for the Iraqi people.

In May 1996, Albright, then U.S. ambassador to the UN, was interviewed on NBC's *60 Minutes*, for an Emmy award–winning segment titled "Punishing Saddam." The first articles on the humanitarian catastrophe caused by the embargo had begun to circulate in the press. Lesley Stahl, who interviewed Albright, echoed those reports: "We have heard that half a million children have died. I mean, that's more children than died in Hiroshima. And, you know, is the price worth it?" Albright replied, "I think this is a very hard choice, but the price, we think the price is worth it."

Obviously, Albright was perfectly well informed about the martyrdom of Iraq's children. UNICEF published the following figures: before the collective punishment implemented by the UN, the infant mortality rate in Iraq was 56 children per 1,000; in 1999, it was 131 per 1,000, with children dying from hunger and lack of medicine—an enormous increase. In eleven years, the massacre led by Albright and implemented by the UN killed several hundred thousand children.

There is no question here of casting doubt on the tyrannical and criminal character of Saddam Hussein's regime. There is no doubt that his regime constituted one of the worst that the Arab world has known. And no doubt, furthermore, that during the eleven years of the embargo, Saddam Hussein, his family, and their accomplices lived like moguls. Year after year, they exported oil illegally through Turkey and Jordan for a total sum estimated at $10 billion. However, the principal responsibility for the death by hunger of hundreds of thousands of Iraqis remains with the sanctions committee of the UN Security Council.

In October 1998, Kofi Annan named Hans Christof Graf von Sponeck UN Assistant Secretary General and UN Humanitarian Coordinator (and thus coordinator of the oil-for-food program) in

Baghdad. His predecessor, an Irishman and thirty-four-year vet-
eran of the UN, Denis Halliday, had just resigned in an uproar. A
historian trained at the University of Tübingen, von Sponeck is the
antithesis of a bureaucrat. During his thirty-seven years of service
at the UN, he had always held field postings, first as an official
with the UN Development Programme in Ghana and in Turkey,
then as a resident representative of the UN in Botswana, India, and
Pakistan. The only post he held far from the front lines in developing countries was that of UNDP regional director in Geneva—
where he was, by his own admission, bored stiff. No one, on the
thirty-eighth floor of the UN headquarters in New York, where the
secretary general, his principal undersecretaries general, and members of their staffs all work, suspected von Sponeck's family history,
which would one day be revealed, with explosive consequences.

In Baghdad, von Sponeck discovered the extent of the humanitarian catastrophe. Like virtually all UN officials and world public opinion, he had previously been totally ignorant of it. As soon
as he understood how the embargo had been hijacked and turned
into a means of collective punishment, and saw the weapon of
hunger in action, von Sponeck spoke out loud and clear to express how appalled he was. He tried to alert the press, his own
government, and above all the Security Council. The Americans
blocked his appearance before the council.

Albright's spokesman, James P. Rubin, tried to discredit von
Sponeck by spreading all kinds of lies about him. "This man in
Baghdad is paid to work, not to speak," he said mockingly at a
press briefing. As for the British ambassador, Stewart Eldon, he
reprimanded von Sponeck, saying, "As the UN Humanitarian
Coordinator you have no business dealing with issues outside your
area of competence! In any case, all you are doing is putting a UN
seal of approval on Iraqi propaganda." Albright finally demanded
von Sponeck's dismissal. Kofi Annan refused. The hatred that
Albright directed against von Sponeck and the campaign of defa-
mation led by Rubin only intensified. But above all, the memory of his father made the situation less and less bearable for von
Sponeck: he could not imagine making himself, on the ground

or from a distance, an accomplice to what not a few were calling genocide. On February 11, 2000, he sent his letter of resignation to New York. Jutta Burghardt, head of the UN World Food Programme in Iraq, did the same. Von Sponeck was succeeded by a dull bureaucrat from Myanmar.

The American bombing of Baghdad on the night of March 7–8, 2003, followed by the ground invasion, officially put an end to the oil-for-food program.

Hans Emil Otto Graf von Sponeck, a General-Leutnant in the Wehrmacht and commander of a division on the Russian front, refused to execute an inhumane order not to retreat that would have resulted in the destruction of his division in the winter of 1941. A military tribunal found him guilty of disobeying a superior officer and condemned him to death. However, an appeal to Hitler for clemency by another general was successful, and the Führer commuted the sentence to six years in the Germersheim Fortress, which was used as a political prison, notably for members of the Norwegian and Danish Resistance.

On July 20, 1944, a group of German officers led by Claus von Stauffenberg attempted to assassinate Hitler at his Wolf's Lair field headquarters near Rastenburg in East Prussia. The attempt, alas, failed. Heinrich Himmler, the head of the SS, vowed to extirpate all organized resistance within the officer corps. Von Sponeck, suspected of anti-Nazi sympathies, was executed by firing squad on Himmler's orders on July 23, 1944.

I asked Hans Emil von Sponeck's son how he had been able to bear for years Albright's crass insults and Rubin's lies, while it must have taken a great deal of strength and courage to break the UN's *omertà*, its own Mafia-like code of silence, in order to stand up in opposition to the powerful sanctions committee, and thereby to renounce his career. Hans Christof Graf von Sponeck is a modest man. He replied, "To have had a father like mine creates certain obligations."

PART V

THE VULTURES OF "GREEN GOLD"

22

A GREAT LIE

There are two principal supply chains for biofuels (also, if rather rarely in English, called agrifuels): bioethanol (a kind of alcohol) and biodiesel. The prefix *bio-*, from the ancient Greek *bios* (human life, the course of human life), which in modern scientific usage refers to all organic life, indicates that the ethanol or diesel fuel is produced from organic matter, or biomass. There is no direct link with the term *bio*, used in many European languages as the equivalent of the English term *organic*, in the sense of "organic agriculture" and "organic food," but the confusion of terms benefits the public image of these fuels, contributing to the impression that they are clean and ecologically sound.

Bioethanol is obtained by the processing of plants containing sucrose (also called saccharose), such as sugar beets and sugarcane, or those containing starch (wheat, corn, and so on). In the first case, the fuel is made by fermenting sugar extracted from the sugar-bearing plants; in the second, by enzymatic hydrolysis of the starch component of cereals (a process in which water splits off macromolecules—in effect, digestion; the fuel obtained by this latter process is also called cellulosic ethanol.) As for biodiesel, it is obtained from vegetable oil or animal fat, transformed by a chemical process called methanolysis, a form of transesterification, in which the fats are reacted catalytically with an alcohol, usually ethanol or methanol.

"Green gold" has been viewed for several years now as a magical and profitable complement to "black gold" (that is, oil). In support of these new fuel sources, the agri-food corporations that dominate the production and trade in biofuels advance an apparently irrefutable argument: the substitution of energy from plant sources for its fossil-fuel siblings is the ultimate weapon in the battle against global climate change and the irreversible degradation of the environment and damage to human life that it will cause.

Here are a few statistics: More than 600 million barrels of bioethanol and biodiesel were produced in 2011. In the same year, some 100 million hectares (247 million acres) were farmed to produce crops for biofuels. With regard to the climate argument, it should be noted that world production of biofuels doubled between 2006 and 2011.

On a global scale, desertification and soil degradation now affect more than a billion people in more than a hundred countries. The dry regions, where an arid or semi-arid climate makes the soil especially vulnerable to degradation, constitute more than 44 percent of the planet's arable land. The consequences of soil degradation are especially serious in Africa, where millions of people depend entirely on the earth to survive as small farmers or pastoralists and there are practically no other means of subsistence. The arid regions of Africa are inhabited by 325 million people (out of nearly 1 billion in the continent as a whole), with heavy concentrations in Nigeria, Ethiopia, South Africa, Morocco, and Algeria, and, in West Africa, south of a line drawn from Dakar to Bamako and Ouagadougou. Currently, around 500 million hectares (1.2 billion acres) of arable land in Africa are affected by soil degradation.

In mountainous countries all over the world, glaciers are retreating. Consider, for example, Bolivia. The highest peak in the country, Nevado Sajama, rises above the high Andean plateau, the Altiplano, to a height of 6,542 meters (21,463 feet); snowcapped Illimani, which overlooks the bowl-like canyon where

La Paz, the capital, lies, reaches 6,450 meters (21,161 feet); and the seracs (pinnacles and ridges of ice on the surface of glaciers) and other glacial features of the Huayna-Potosí in the Cordillera Real reach 6,088 meters (19,973 feet). The snows of these peaks glisten in the sun- and moonlight. The inhabitants of the *ayllus* (traditional indigenous Andean communities, dating back to pre-Inca times, which have become newly visible under President Evo Morales) and their priests consider the mountains sacred and eternal. But eternal they are not.

For global warming is melting the glaciers and making the snowfields retreat. The rivers are swelling. The situation is becoming catastrophic, especially in the Yungas forest on the eastern slopes of the Andes, where torrential floods caused by melting snow tear through the villages on the riverbanks, killing livestock and people, destroying bridges, and gouging out ravines. And, eventually, the loss of glacial volume may pose critical problems for water resources.

Everywhere in the world, the deserts are growing. In China and Mongolia, on the edges of the Gobi Desert, every year more and more pasture and cropland is swallowed up by the sand dunes as they advance upon the fertile land. In the Sahel, the Sahara is advancing in some zones by five kilometers (three miles) a year. I have seen in Mek'ele, in the Tigray region of northern Ethiopia, skeletal women and children trying to survive on land that erosion has transformed into acres of dust. Stalks of teff, the national cereal, barely grow 30 centimeters (1 foot) tall, as opposed to 1.5 meters (4 feet) in Gondar or Sidamo.

The destruction of ecosystems and the degradation of vast agricultural zones worldwide, but above all in Africa, is a tragedy for small farmers and pastoralists. The UN estimates that there are 25 million "ecological refugees" or "environmental migrants," that is, human beings who are forced to leave their homes following natural disasters such as floods, droughts, and desertification, to end up having to fight to survive in the shantytowns of the big cities. Soil degradation provokes conflicts, especially between farmers and pastoralists. Many conflicts, especially in sub-Saharan Africa, including the Darfur region of Sudan, are closely linked to the phenomena of drought and desertification, which, as they worsen, spark conflicts between nomads and sedentary farmers for access to resources.

The global corporations that produce biofuels have, however, succeeded in persuading the majority of world public opinion and virtually all of the Western nations that energy from plant sources constitutes the miracle weapon against climate change. Yet their argument is a lie. It ignores the methods and environmental costs of biofuel production, which requires both water and energy.

All over the planet, drinking water is becoming scarcer and scarcer. One person in three is currently reduced to drinking polluted water; nine thousand children under age ten die each day from ingesting water unfit for drinking. Of the 4 billion cases of diarrhea counted annually, 2.2 million are fatal, mainly among infants and children. But diarrhea is only one of numerous illnesses transmitted by poor-quality water: others include trachoma, schistosomiasis (bilharzia), cholera, typhoid, dysentery, hepatitis, malaria, and others. Many of these diseases are caused by the presence of pathogenic organisms in the water (bacteria, viruses, worms). According to WHO, up to 80 percent of diseases and more than one-third of deaths in developing countries are, at least in part, attributable to the consumption of contaminated water.

According, again, to WHO, one-third of the world's population still has no access to safe drinking water at an affordable price, and half of the world's population has no access to sanitation and sewage facilities. Around 285 million people live in sub-Saharan Africa without regular access to unpolluted water; 248 million in South Asia likewise; 398 million in East Asia; 180 million in Southeast Asia and the Pacific; 92 million in Latin America and the Caribbean; and 67 million in the Arab countries. And it is, of course, the most impoverished who suffer the most from the lack of water.

From the point of view of the planet's water reserves, the production, every year, of tens of billions of liters of biofuels constitutes a veritable catastrophe. It takes, in fact, four thousand liters of water to produce one liter of bioethanol. And it is not Noël Mamère or some other reputedly "doctrinaire" ecologist who makes this claim, but Peter Brabeck-Letmathe, president of the biggest food corporation in the world, Nestlé.

In addition, a detailed study by the Paris-based Organisation for Economic Co-operation and Development, whose membership comprises most of the industrialized nations, has calculated that the amount of fossil fuel needed to produce one liter of biofuel is very considerable indeed. And as the *New York Times* has noted soberly, given the great quantity of energy required in their production, "a large-scale effort across the world to grow crops for biofuels would add carbon dioxide to the atmosphere rather than reduce it."

23

BARACK OBAMA'S OBSESSION

By far the most powerful producers of biofuels in the world are U.S.-based multinational corporations. Each year, they receive many billions of dollars in government aid. President Barack Obama has discussed biofuels as part of a critically important national "clean energy" strategy. In his State of the Union speech in 2011, Obama said, "With more research and incentives, we can break our dependence on oil with biofuels"—clearly, the bioethanol and biodiesel programs constitute an essential national security priority.

In 2011, subsidized with $6 billion in public funds, the American biofuel companies will have burned 38.3 percent of the national corn harvest, up from 30.7 percent in 2008. And since 2008, the price of corn on the world market has risen by 48 percent. (In 2008, the American companies burned 138 million tons of corn, which equals 15 percent of world consumption.)

The United States is also by far the most dynamic and most important industrial power on the planet. Despite a relatively small population of about 313 million, compared to more than 1.34 billion in China and 1.17 billion in India, the United States produces slightly more than 25 percent of all the industrial goods manufactured annually on the planet. The primary material of this impressive machine is oil. The United States burns on average

20 million barrels of oil per day, or about one-quarter of world production; 61 percent of this volume, or a little more than 12 million barrels a day, is imported. Only 8 million barrels are produced in the United States, in Texas, offshore in the Gulf of Mexico, and in Alaska.

For the American president, this dependence on foreign oil is obviously a preoccupation. And even more worrying, the bulk of this imported oil comes from parts of the world where political instability is endemic, where Americans are not liked, and, in short, where production and export to the United States are not guaranteed. As a consequence of this dependence, the government in Washington must maintain in these regions, especially in the Middle East, the Persian Gulf, and Central Asia, a very expensive military presence, on land, at sea, and in the air.

In 2009, for the first time, expenditures on armaments by UN member nations (beyond budget allocations for maintenance of their militaries per se) exceeded $1 trillion. Of this amount, the United States by itself accounted for 41 percent of the total (China, with the world's second-largest military, accounted for 11 percent). American taxpayers also pay for $3 billion in military aid to Israel annually. Furthermore, they finance very expensive military bases in Saudi Arabia, Kuwait, Barhrain, and Qatar. Despite the magnificent Egyptian people's revolution in 2011, Egypt remains an American protectorate. And U.S. taxpayers send $1.3 billion to the generals in Cairo. It must be understood, moreover, that if President Obama wants to have the slightest chance of funding his social programs, especially reform of the U.S. health care system, he must, urgently and massively, reduce the budget of the Pentagon. Yet this budgetary reduction is possible only by substituting, as much as possible, fuel from plant sources (produced in the United States) for fossil fuels (mostly imported).

George W. Bush was the initiator of the biofuels program. In January 2007, he set the following goals: in ten years, the United States should reduce by 20 percent its consumption of fossil fuels and increase by a factor of seven its production of biofuels, from a

2007 level of production of slightly less than about 18 billion liters (nearly 5 billion gallons).

Burning millions of tons of food on a planet where a child under age ten dies of hunger every five seconds is obviously appalling. Spokespersons for the agri-food corporations try to defend the industry against criticism. They do not deny that it is morally questionable to divert food from its primary purpose and to use it as raw material for fuel production. But we have no cause to worry, they promise us. Soon there will be a "second generation" of biofuels produced from agricultural by-products, wood shavings, and plants such as species of the *Jatropha* genus that grow only on arid land (where no production of food crops is possible). And already, they add, production techniques allow the corn stalk to be processed for fuel without damaging the grain. Perhaps, but at what price?

The word *generation* evokes biology, suggesting a logical and necessary progression. But such terminology is, in this instance, very deceptive. Because if the possibility of so-called second-generation biofuels does indeed exist, their production is clearly going to be more expensive than the first generation's, owing to the need to separate out the fuel-producing components of the plants involved rather than using the entire plant, as well as other intermediate steps required in processing. And as a result, in a market dominated by the principle of profit maximization, these second-generation biofuels will play only a marginal role.

The gas tank of a midsize car burning bioethanol holds 50 liters (about 13 gallons). To produce that much bioethanol, 358 kilograms (789 pounds) of corn must be destroyed. In Mexico, in Zambia, corn is the staple food. A Mexican or Zambian child could live off 358 kilograms of corn for a year.

In my view, to invoke a phrase widely used by diverse critics of the industry, the question of biofuels can be summed up as this choice: full gas tanks or empty stomachs.

24

THE CURSE OF SUGARCANE

Not only do biofuels devour every year hundreds of millions of tons of corn, wheat, and other foodstuffs; not only does their production send millions of tons of carbon dioxide into the atmosphere; but, in addition, biofuel production causes social disasters in the countries where the global corporations that produce them become dominant. Consider the example of Brazil.

The jeep advances with difficulty over the rutted roadway that winds through the Capibaribe valley. The heat is suffocating. The green ocean of sugarcane stretches away to infinity. James Thorlby is sitting in the front seat, beside the driver. We are advancing into enemy territory. In the valley, many *engenhos* (sugarcane plantations) are being occupied by members of the MST, the Landless Rural Workers' Movement. The sugar barons have close ties to the military police, which are in effect a federal police force. Not to mention the death squads, the plantation gunmen, who prowl throughout the region.

Thorlby is Scottish and a priest. From Bahia to Piauí, all over the northeast, he is known as Padre Tiago (from Santiago, the Portuguese equivalent of James). His friend Chico Mendez has been assassinated. Thorlby is still alive. For the time being at least, he adds. Tiago has a macabre sense of humor: "I prefer to sit in front. The gunmen are superstitious. . . . They are more afraid to shoot at a priest than a socialist from Geneva." However, so far only swarms of mosquitoes have attacked us.

The sun sets red behind the horizon as, at last, we arrive within sight of the plantation. Hiding the vehicle in the bushes, we continue on foot, Thorlby, a union organizer, Sally-Anne Way, Christophe Golay, and I. The little adobe houses where the cane cutters live with their families, all painted blue, line both sides of a muddy rivulet. The entrance to each house is elevated: you have to climb three steps to get to the little stone terrace on which the house is built. The system is clever: it protects the inhabitants from rats and from sudden floods of the nearby stream.

The children—some *caboclo* (of mixed European and indigenous ancestry), some Afro-Brazilian, and some with more pronounced indigenous facial features—are happy despite a degree of undernutrition that we can see immediately in the thinness of their arms and legs. Many of them have the belly swollen by worms and the thin, reddish hair that are symptoms of kwashiorkor. The women are poorly dressed, with ebony-colored hair framing bony faces with hard eyes. Few of the men have all their teeth. Tobacco has stained their hands deep yellow.

Colored hammocks hang crisscross beneath the beamed ceilings. Parrots perch in their cages beneath the eaves. Behind the houses donkeys bray. Brown goats gambol in the meadows of sparse grass. The smell of roasted corn fills the air. The mosquitoes make a dull noise, like distant bombers.

The struggle of the workers of the Trapiche *engenho* is a good example of the overall struggle. The vast fields that disappear in the evening mist were formerly state-owned land, or *terra da União*. The land now controlled by the plantation consisted, only a few years ago, of subsistence farms comprising one or two hectares (2.5 to 5 acres). The families who worked these farms were poor, but they lived in safety, with a certain level of well-being and in relative freedom.

With a considerable amount of capital at their disposal, and because they enjoyed excellent relations with the federal government in Brasília, financiers were able to get the relevant authorities to "downgrade" the land, that is, to privatize it. The small farmers who lived here, growing beans and cereals, were forced off the land and driven to the shantytowns of Recife, except for those who agreed, for wretched wages, to become cane cutters. Today, the cane cutters are severely exploited.

A long judicial process undertaken by the MST against the new proprietors had ended with a decision in favor of the landowners shortly before our visit. The local judges were no more impervious than the politicians in Brasília to the financial advantages—for them—of the privatization of public lands.

In Brazil, the program of biofuel production enjoys absolute priority. And sugarcane constitutes one of the most profitable raw materials for the production of bioethanol. The Brazilian program that aims to accelerate the rate of bioethanol production has an odd name: Pro-alcohol (or the National Alcohol Program). It is a source of pride for the government. In 2009, Brazil consumed 14 billion liters (3.7 billion gallons) of bioethanol and biodiesel and exported 4 billion liters (1.06 billion gallons). The government's dream: to export as much as 200 billion liters (52.8 billion gallons) per year.

The state energy company, Petrobras, is dredging new deep-water harbors at Santos, in São Paolo state, and in Guanabara Bay, in Rio de Janeiro state. Over the next ten years, Petrobras will invest $85 billion in the construction of new port facilities. The federal government in Brasília wants to increase the amount of land under cultivation for sugarcane to 26 million hectares (more than 64 million acres). Against the bioethanol giants, the toothless cane cutters of the Trapiche plantation haven't got a chance.

The implementation of the Brazilian Pro-alcohol plan has led to the rapid concentration of land ownership in the hands of a few Brazilian sugarcane barons and multinational companies. The biggest sugar-producing region in the state of São Paolo is Ribeirão Preto. Between 1977 and 1980, the average size of the properties there increased from 242 to 347 hectares (from 598 to 860 acres). The concentration of land ownership, and thus of economic power, in the hands of a few big companies and plantations has rapidly become widespread, and the process has accelerated since 2002. This trend toward concentrated land holding obviously works to the detriment of small and medium-size family farms.

One FAO expert has written that the size of an average plantation in the state of São Paolo, which was 8,000 hectares (19,768 acres) in 1970, had increased to 12,000 hectares (29,653 acres) by 2008. Plantations comprising 12,000 acres or more in 1970 had reached an average size of 39,000 hectares (96,371 acres) or more by 2008. Plantations of 40,000 or 50,000 hectares (up to 123,553 acres) were (and are) not rare. Conversely, the average area of plantations under 1,000 hectares (2,471 acres) had fallen by 2008 to 476 hectares (1,176 acres). The concentration of land in the state of São Paolo not only is the result of the buying and selling of land but is frequently also caused by formerly independent farmers being forced by the big landowners to rent their land to the big plantations.

This reorientation of agriculture toward a model of monopoly capitalism has left behind on the sidelines those who have not had the means to buy farm machinery, agricultural inputs, and land, and thereby to embark on intensive sugarcane cultivation. These farmers excluded from the emerging model have endured great pressure to agree to rent or sell their land to big neighboring estates. The period from 1985 to 1996 saw no fewer than 5.4 million small farmers evicted from their land and the disappearance of 941,111 small and medium-size farms across Brazil.

The monopolization exacerbates inequality and increases rural poverty (as well as urban poverty caused by the effect of the rural exodus). Moreover, the exclusion of smallholders endangers the country's food security, since it is the small farms that guarantee food production. As for rural families headed by a woman, they have less easy access to land and suffer from increased discrimination.

In short, the development of "green gold" production on the agro-exporter model immensely enriches the sugar barons but little by little weakens small farmers, tenant farmers, and the *bóias-frias*. The program signs the death warrant for the small and medium-size family farm—and thereby for the country's food sovereignty.

Together with the Brazilian sugar barons, the Pro-alcohol program obviously profits the big foreign global corporations such as Louis Dreyfus, Bunge, the Noble Group, and Archer Daniels Midland, as well as hedge funds run by such figures as George Soros and the sovereign funds of China. According to a report from the NGO Ethical Sugar, China and the state of Bahia have signed an agreement permitting China to open twenty ethanol factories in Recôncavo, a vast sugar-producing region stretching inland from the Bay of All Saints in Bahia, between 2011 and 2013. In a country such as Brazil, where millions of people are demanding the recognition of their right to own their own small plot of land, where food security is threatened, the monopolization of land by multinational corporations and sovereign funds constitutes an egregious scandal.

At the UNHRC and before the UN General Assembly I fought against the Pro-alcohol plan. Opposing me was Paulo Vanucci, a friend, a former guerilla of the VAR-Palmares, and a hero of the resistance against the Brazilian dictatorship. He was truly sorry. Even President da Silva, during his appearance before the council in 2007, attacked me specifically by name from the rostrum. Vanucci and Lula had one killer argument in their arsenal: "Why worry about the expansion of the ocean of sugarcane? Ziegler is the Special Rapporteur on the Right to Food. But the Pro-alcohol plan has nothing to do with food. Sugarcane is not edible. Unlike the Americans, we Brazilians are burning neither corn nor wheat."

This argument is unacceptable, because the agricultural boundaries in Brazil are being permanently displaced: sugarcane is advancing toward the interior of the Brazilian Highlands, the country's heartland, located on the continental plateau, forcing herds of livestock that have grazed there for centuries to move west and north. In order to create new grazing lands, the plantations and directors of the global corporations that raise cattle are burning the forests—by tens of millions of hectares each year.

This destruction is irreparable. The soils of virgin forests in the Amazon Basin and the Mato Grosso in western Brazil have only a thin layer of humus. Even in the unlikely event that the government leaders in Brasília were to be seized by a sudden spell of clear-sightedness, they could never re-create the Amazonian rainforest, justly known as the "planet's lungs" because it plays an essential role in regulating not only the region's precipitation but the climate of the entire planet. According to an estimate by the World Bank, at the current rate of slash-and-burn destruction of the forest to create rangeland, 40 percent of the Amazon rainforest will have disappeared by 2050.

To the extent that Brazil has increasingly substituted cultivation of sugarcane for food crops, it has joined the vicious circle of the international market in food: compelled to import the food products that it no longer produces itself, the country thereby increases world demand—which in turn leads to increased prices. The food insecurity in which a great part of the Brazilian population lives is thus directly linked to the Pro-alcohol program. The program particularly affects the sugarcane-growing regions, because there the consumption of staple foods depends almost entirely on the purchase of imported food products subject to wide fluctuations in prices. As David and Marcia Pimentel point out, many smallholders and agricultural workers are net buyers of food products because they do not own a sufficient amount of land to produce enough food for their families. This is why, in 2008, many Brazilian farmers and their families went hungry: they could not afford to buy enough food owing to the brutal explosion in food prices.

In the sugarcane fields of Brazil there still exist many practices that approach the enslavement seen in the period before 1888, when slavery was finally abolished in the country. Cutting cane is extraordinarily hard work. Cutters are paid by the job. Their only tool is a machete—and, if the foreman has a heart, they also wear leather gloves to protect their hands from abrasions. The minimum wage is rarely honored in the countryside.

And yet, owing to the Pro-alcohol program, the army of those laboring under the curse of sugarcane ceaselessly grows. Along with their families, the cane cutters migrate from one harvest to the next, from one plantation to the next. The sedentary cane cutters of the Trapiche *engenho* are today an exception. The multinational corporations also prefer to employ migrant workers. They thereby save on obligatory contributions to social welfare programs and reduce their costs of production. This practice has severe social and human costs.

Anxious to reduce their costs, the producers of biofuels exploit migrant workers by the millions, according to an ultra-pro-free-market neoliberal capitalist model of agriculture: low wages, inhumane work hours, nearly nonexistent accommodations, and work conditions that approach slavery. These conditions have disastrous consequences for the health of the workers and their families. This is why the cutters, and even more their wives and children, often die from tuberculosis and undernutrition.

In Brazil there are 4.8 million rural workers *sem terra*. Many of these landless workers are on the road, without permanent homes, selling their labor according to the seasons. Those who live in villages or country towns, or in shacks on the edges of the big plantations, at least have access to a minimum of social services.

The transformation of vast regions into zones of sugarcane monoculture "casualizes" labor and makes employment insecure owing to the seasonal nature of the cane harvest. Once the harvest is done in the south, the workers must travel 2,000 kilometers (1,200 miles) from there to the northeast, where the seasons are reversed. In this way, they travel to find work every six months, covering immense distances. Far from their families, they are uprooted, more vulnerable than if they could work at home. The *bóias-frias*, who are not migratory, are no better off, never knowing how long they will be employed—a day, a week, a month?

Such vulnerability, such mobility make it even harder for these workers to defend the few rights they do have. The cane workers are in general unable to report the frequent abuses committed by their employers. Moreover, the legislation that should protect them is almost nonexistent:

> Many live and suffer much as their ancestors did—as slaves on sugar plantations. Government investigators occasionally liberate a handful of cane workers, but in such a big

country the officials are few and far between. The real
power lies in the hands of militias, or capangas, working for
the sugar barons. . . . [A]nyone who makes trouble quickly
finds himself face-to-face with the capangas, who crisscross
the plantations in Jeeps and on dirt bikes. They carry radios
and weapons. Officially, they are considered security guards
who watch over the plantations. In reality, the capangas circle the workers like aggressive dogs encircling a herd.

Few women work in the cane fields, because it is very difficult
for them to achieve the fixed targets of cutting ten or twelve tons
of cane per day. However, according to the FAO, women who
work on a seasonal basis or as day laborers, "due to existing social
inequalities tend to be particularly disadvantaged, compared to
men, in terms of wages, working conditions and benefits, training
and exposure to safety and health risks." Thousands of children
work on the plantations as well. In 2004, it was estimated that
2.4 million workers under age seventeen labored in Brazilian agriculture, including 22,876 on sugarcane plantations.

Brazilian cane cutters' income is also threatened by mechani-
zation. Although they are unusable in the rugged terrain of the
northeast, mechanical sugarcane harvesters are replacing manual
labor on many plantations in the state of São Paulo, the country's
leading sugar-producing state; in 2010, 45 percent of the harvest
there was mechanized.

Gilberto Freyre's famous book, *The Masters and the Slaves: A Study
in the Development of Brazilian Civilization*, to which I have already
referred, is a denunciation of the curse of sugarcane.

The story Freyre recounts dates back to the early days of the
Portuguese colony, when Tomé de Souza, Brazil's first governor-
general, sailed his caravel into the Bay of All Saints on March 29,
1549, and not long after established Salvador as the capital of the
colony in Bahia. By the seventeenth century, sugarcane had inun-
dated first the Recôncavo region, then the Capibaribe River valley in Pernambuco, and finally the coastal zones and all of rural

Sergipe and Alagoas. The sugar industry was at the foundation of the slavery-based economy. The *engenhos* were sheer hell for the slaves, but they constituted a source of phenomenal wealth for their masters.

The sugar monoculture ruined Brazil. Today, it has returned. Once again, the curse of sugarcane has descended upon Brazil.

POSTSCRIPT: HELL IN GUJARAT

The slavery-like conditions in which cane cutters work are not unique to Brazil. Thousands of migrant cutters in many other countries endure the same kind of exploitation.

The Bardoli Sugar Factory plantation in the Surat region of Gujarat, India, supplies the largest sugar factory in Asia. The vast majority of the men who work in the cane fields there are members of various Adivasi, indigenous tribal groups, famous for their baskets and furniture woven from reeds.

Living conditions on the plantation are horrifying: the food provided by the factory bosses is infested with worms, and there is lack of clean water as well as of wood for cooking. The Adivasi and their families live in shacks made from branches open to scorpions, snakes, rats, and wild dogs.

The irony of the situation is that, for fiscal reasons, the Bardoli Sugar Factory is registered as a cooperative. In India, one of the most restrictive laws is the one that regulates the obligations and public oversight of cooperatives, the Cooperative Societies Act of 1912, under which specific officials are appointed to oversee cooperatives. But the cane cutters never see those officials. The state government of Gujarat ignores their suffering.

Why don't the cutters appeal to the courts? The Adivasi are much too afraid of the *mukadam*s, the labor contractors who hire the cutters for work on the plantation. The level of unemployment in Gujarat is so high that a cutter who makes the least complaint will be replaced within the hour by a more compliant worker.

25

CRIMINAL RECOLONIZATION

During the sixteenth session of the UNHRC in March 2011, La Via Campesina, together with two other NGOs, the FoodFirst Information and Action Network (FIAN) and the Geneva-based Centre Europe–Tiers Monde, organized a side event, an informal consultation on the protection of farmers' rights, such as the rights to land, seed, water, and so on. The dauntless Pitso Montwedi, Chief Director for Human Rights and Humanitarian Affairs in South Africa's Department of International Relations and Cooperation in Pretoria, declared on this occasion, "First they took our people, then they took our land . . . we are living through the recolonization of Africa."

The curse of "green gold" is in effect spreading today to many countries in Asia, Latin America, and Africa. (Brazil is the main seller of equipment for the production of biofuels.) Almost everywhere in the world, but above all in Asia and Latin America, the monopolization of land by biofuel corporations is accompanied by violence. Colombia provides the paradigmatic example.

Colombia is the world's fifth-largest producer of palm oil: 36 percent of the oil produced is exported, mainly to Europe. In 2005, 275,000 hectares (679,500 acres) were devoted to oil palm cultivation. Palm oil is used in the production of biofuel. One hectare (2.5 acres) produces about 5,000 liters (1,321 gallons) of biodiesel.

In practically every region of Colombia where oil palm has been planted, human rights violations have accompanied the planting of the trees: illegal appropriation of land, forced displacement of communities, targeted assassinations, disappearances. The scenario, repeated in almost every region affected, begins with the forced displacement of the local population, achieved by the "pacification" of the zone by paramilitary units in the pay of private global corporations. Between 2002 and 2007, 13,634 people, including 1,314 women and 719 children, were killed or disappeared essentially as a result of attacks by paramilitaries.

Here is just one example: In 1993, the Colombian government recognized, by passing its Ley 70 (Law Number 70), the property rights of the African-Colombian people who traditionally farm the land of the Curvaradó and Jiguamiandó river basins. The law stipulates that no one may acquire any substantial section of the 150,000 hectares (371,000 acres) encompassed by the two river basins without the consent of representatives of the communities affected. But the reality on the ground is entirely different. The farmers and their families have fled the paramilitaries. The global palm oil corporations plant their trees in peace.

The paramilitaries arrived in the region in 1997, leaving desolation in their wake: houses burned down, targeted assassinations, threats, massacres. Human rights organizations have documented between 120 and 150 assassinations and the forced displacement of 1,500 people. As soon as the people had been driven off their land, the corporations began to plant the first palms. In 2004, 93 percent of the common land of the communities in the region was occupied by oil palm plantations.

Here is another example: the long battle finally lost by the farming families of Las Pavas, which has been described by journalist Sergio Ferrari. There, the godfathers of organized crime joined with the plantation owners to dispossess a community of more than six hundred families of their land in the department of Bolívar in northern Colombia. The tragedy goes back to the 1970s, when the farmers were forced out by plantation owners who sold their parcels to Jesús Emilio Escobar, an uncle of the drug lord Pablo Escobar. In 1997, Escobar abandoned the property and the community took back its land, where they cultivated rice, corn, and bananas. But six years later the farmers were once again evicted by the paramilitaries.

The courageous farmers of Las Pavas refused to merely stagnate in their displaced persons camp. Little by little, their families returned to Las Pavas. In 2006, they presented to the Ministry of Agriculture a demand for the recognition of their property rights. This was the moment that Escobar chose to dislodge the farmers' families by force once again, destroying their harvests and selling their land to the El Labrador Consortium (a joint venture of two companies, Aportes San Isidro and Tequendama), which specializes in the large-scale cultivation of oil palm.

In July 2009, the farmers, who continued to cultivate part of their lands despite threats, were one more evicted, this time by the police, an action that even the minister of agriculture himself deemed illegal. In 2011, a new president was in power in Bogotá, Juan Manuel Santos. His predecessor, Alvaro Uribe, was affiliated with the paramilitary death squads. Santos, for his part, is close to the social circles that include the plantation owners. The directors of the palm oil companies, in particular those who run Tequendama, are his friends. The farmers of Las Pavas and their families have not the slightest chance of obtaining justice.

Consider what is happening in another part of the world, Africa. In Angola, the government has announced projects designating 500,000 hectares (1,235,500 acres) for the cultivation of biofuel crops. The effects of these projects will coincide with the massive expansion of the banana and rice monocultures led by the multinationals Chiquita and Lonrho, as well as by certain Chinese companies. In 2009, Biocom (Companhia de Bioenergia de Angola) began to plan sugarcane on a 30,000-hectare (74,100-acre) site. Biocom is partners with the Brazilian Odebrecht conglomerate and the Angolan companies Damer and Sonangol (the Angolan state petroleum company).

The Portuguese firm Quifel Natural Resources plans for

its part to cultivate sunflower, soy, and jatropha in the equatorial province of Cunene. The company plans to export the harvests to Europe, where they can be transformed into biofuels. The Portuguese company Gleinol has produced agrodiesel on 13,000 hectares (32,100 acres) since 2009. Sonangol, in association with the Italian petroleum consortium ENI, plans to expand its existing plantations of oil palm in the province of Kwanza-Norte to produce biofuels.

In Cameroon, Socapalm (Société Camerounaise de Palmeraies; Cameroon Palm Plantations Company), formerly a state-owned enterprise, is today partially owned by the French group Bolloré. Socapalm has announced its intention to expand palm oil production. The company owns palm plantations in the country's central, southern, and coastal regions. In 2000, it signed a sixty-year lease on 58,000 hectares (143,300 acres). Bolloré, in addition, directly owns the Sacafam (Societé Agricole Forestière du Cameroon; Cameroon Agricultural and Forestry Company) plantation, with 8,800 hectares (21,700 acres).

In Cameroon, the oil palm plantations destroy virgin forests in the Congo River basin, aggravating further the long-term process of deforestation caused by the combined effects of forestry and clearing for agriculture. The federal government has since the 1960s supported the development of the palm oil industry through its state companies, Socapalm, the Cameroon Development Corporation, and the Compagnie des Oléagineux du Cameroun (Cameroon Oilseed Company). Yet the tropical forest of the Congo basin is the world's second-largest after the Amazon and constitutes one of the planet's principal carbon sinks (natural features, such as forests, oceans, peat bogs, and prairies, that absorb carbon dioxide from the atmosphere via photosynthesis, storing part of the absorbed carbon and returning oxygen to the atmosphere). It is important to understand also that numerous hunter-gatherer communities depend upon this forest and its rich biodiversity for their survival. As a result, these communities are risk of disappearing.

The government of Benin has proposed the conversion of

300,000 to 400,000 hectares (741,300 to 988,400 acres) of wetland areas into oil palm plantations in the southern part of the country. The oil palm is in fact a plant that originated in wetlands, but the palm plantations are going to drain the wetlands and destroy the rich biodiversity that they shelter.

But it is in the Democratic Republic of the Congo (DRC) that some of the biggest biofuel projects of all are under way. In July 2009, the Chinese company ZTE Agribusiness announced its intention to develop an oil palm plantation of 1 million hectares (2,471,000 acres) for the purpose of producing biofuels. ZTE had previously, in 2007, announced that it would invest up to $1 billion in a new plantation covering 3 million hectares (7,413,000 acres). The Italian multinational company ENI, with 70,000 hectares (173,000 acres), also has a large oil palm plantation in the DRC.

The Marxist Ethiopian government is also launching enthusiastically into the alienation of its land. It has put nearly 1.6 million hectares (nearly 4 million acres) up for grabs for investors eager to develop sugarcane and palm oil plantations. By July 2009, 8,420 local and foreign investors had received the necessary authorizations to break ground.

In 2007, the Japanese company Biwako Bio-Laboratory was farming 30,000 hectares (74,100 acres) of *Jatropha curcas* in Kenya, with the goal of producing jatropha oil, and plans to expand its plantings to 100,000 hectares (247,100 acres) within ten years. Another company in Kenya, the Belgian corporation HG Consulting, provides financing for the Ngima project, which uses sugarcane grown by small farmers under contract working a total of 42,000 hectares (103,800 acres). The Canadian company Bedford Biofuels has acquired 160,000 hectares (395,300 acres) to plant jatropha, with an option for 200,000 hectares (494,200 acres) more.

In 2008, Marc Ravalomanana, the president of Madagascar (2002–9), concluded a secret agreement with the South Korean multinational Daewoo conglomerate agreeing to transfer 1 million hectares (2,471,000 acres) of arable land to the company's control. Daewoo would be granted this concession for ninety-nine years entirely free of charge. Daewoo planned to plant oil palm to produce ethanol. The company's only obligation was to build roads, irrigation canals, and warehouses. On November 19, 2008, the *Financial Times* of London revealed the contents of the agreement. Ravalomanana was driven out of office by his enraged people. His successor canceled the contract.

Sierra Leone is the poorest country in the world. Addax Bioenergy, a private global company based in Lausanne, recently acquired a concession of 20,000 hectares (49,400 acres) of fertile land there. Addax wants to plant sugarcane to produce bioethanol for the European market. Addax belongs to Jean-Claude Gandur, a Swiss multibillionaire born in Azerbaijan, who made a colossal fortune in the petroleum industry. Joan Baxter visited the site of Addax's plantation in Sierra Leone, reporting:

> Spread out among twenty-five villages in central Sierra Leone, small farmers produce their own seed and cultivate rice, manioc, and vegetables. Adama, who is planting manioc, tells me that the revenues she earns from her harvest will enable her to take care of the needs of her paralyzed husband and to pay the school fees for her three children. Charles, who returns home from the fields in the heat of late afternoon, will be able to send his three little kids to school thanks to what his small farm produces.
>
> Next year, most of these farmers will not be able to cultivate their land. . . .
>
> Adama does not yet know that she is soon going to lose the fields of manioc and pepper that she farms on the highlands. Gandur signed his contract with the government in Freetown. The farmers living in the twenty-five villages only heard about their impending ruin from hearsay.

The problem is common throughout sub-Saharan Africa. For rural land, there generally is no system of land registry; for land in urban areas, such systems exist only in a few cities. In theory, all land belongs to the state. Rural communities have only a right of usufruct to the land they occupy, a right of use and enjoyment but not of ownership.

Gandur is a well-informed capitalist. He takes no risks. He got his project in Sierra Leone financed by the European Investment Bank and the African Development Bank. In Sierra Leone, as in numerous other countries in the southern hemisphere, these two banks (like others elsewhere) function as active accomplices in the destruction of African farming families' way of life of. Three supplementary concessions are under negotiation between the government and Addax—again, with the support of the two public banks. These new concessions concern lands where gigantic oil palm plantations will be developed.

Sierra Leone is emerging from eleven years of horrifying civil war. Despite the end of combat in 2002, reconstruction is not progressing. Nearly 80 percent of the population lives in extreme poverty, seriously and permanently undernourished.

Addax's feasibility study plans for the importation of machinery, trucks, and herbicide sprayers, and the use of chemical fertilizers, pesticides, and fungicides. Addax chose the land it did for a specific reason: it is bounded by one of Sierra Leone's most important rivers, the Rokel. The contract includes no clause limiting the amount of water that Addax will be permitted to pump from the river to irrigate its plantations, nor specifying the uses to which wastewater may be put. The farmers of the entire region are threatened with a lack of water for drinking and for irrigation and with the danger of water pollution.

Formally, Gandur signed a contract renting the land for fifty years, for the cost of 1 euro per hectare (about 42 cents per acre). The contract promises Addax exemptions from personal income taxes and customs duties on imported materiel. Gandur is clever. He has linked his business to an influential local businessman with a long career in mining and oil, Vincent Kanu, as well as with Ibrahim Martin Bangura, the member of parliament for the district. On paper, Sierra Leone is a democracy. In fact, MPs reign over their constituencies like satraps in the ancient Persian Empire. Gandur gave Bangura the task of "explaining" details of the project to the local people. According to Bangura, the dispossessed farmers will benefit, by way of compensation, from the four thousand jobs that Addax has promised to create. But an independent field study has proven that this promise is false. Few jobs are planned. Moreover, we might ask, jobs under what conditions? No one has said. There is, however, one indication. As of 2011, Addax was employing about fifty people to watch over the young shoots of sugarcane and manioc planted on the shores of the Rokel River. Addax paid them a daily wage of 10,000 leones, or about $2.37.

Gandur's deal in Sierra Leone is typical of most of the acquisitions of land by the lords of green gold. And the corruption of local associates obviously plays a key role in the tactics of dispossession of local farmers. Adding to the scandal is that taxpayer-funded public banks such as the World Bank, the European Investment Bank, and the African Development Bank finance the confiscation of land.

What will become of Adama and Charles, their children, their extended families, their neighbors? They will be driven off their land. Where will they go? To the sordid shantytowns of Freetown, seething with rats, where children sell their bodies and their fathers waste away in permanent unemployment and despair.

Biofuels are catastrophic for society and the global climate. Their production reduces the amount of land available for food crops, destroys family farms, and increases world hunger. It sends great quantities of carbon dioxide into the atmosphere and sucks up an enormous volume of drinking water.

There is no doubt that the consumption of fossil fuels must be rapidly and massively reduced. However, the solution lies not in biofuels but rather in the reduction of energy consumption and in alternative sources of clean energy such as wind and solar.

Bertrand Piccard is one of the most radiantly optimistic men I know. From March 1 to 21, 1999, together with Brian Jones, he completed the first nonstop trip around the globe in a balloon. Today he is preparing to be the first to circle the Earth in a

100 percent solar-powered, piloted, fixed-wing aircraft, the *Solar Impulse*. Piccard once told me, smiling, "I want to contribute to liberating humanity from petroleum."

In 2007, before the UN General Assembly in New York, I declared, "To produce biofuels with food is criminal." I demanded that the practice be forbidden. The vultures of green gold reacted forcefully. The Canadian Renewable Fuels Association, the European Bioethanol Fuel Association, and the Brazilian Sugarcane Industry Association, three of the most powerful federations of bioethanol producers, sent representatives to Kofi Annan to denounce my declaration as "apocalyptic" and "absurd."

I have not changed my mind.

On a planet where a child under age ten dies of hunger every five minutes, to hijack land used to grow food crops and to burn food for fuel constitutes a crime against humanity.

PART VI

THE SPECULATORS

26

THE "TIGER SHARKS"

The tiger shark is a very large member of the family of carcha-rhinid sharks, and an extremely voracious carnivore. With its big teeth and black eyes, it is one of the most feared animals on the planet. It inhabits tropical and temperate oceans worldwide, with a preference for hunting in murky coastal waters. With its powerful jaws, the tiger shark can exert a pressure of several tons per square inch. Like most pelagic (or open-ocean) sharks, the tiger spends most of its life in motion, and is able to detect a tiny amount of blood in an enormous volume of water and follow a trail of blood to its source over a distance of at least 0.4 kilometer (a quarter mile).

The speculator in food commodities working on the Chicago Commodity Stock Exchange corresponds rather well to the de-scription of the tiger shark. He too is capable of detecting his vic-tims across great distances and of killing them in an instant, all while satisfying his voracious appetite—or, in other words, real-izing colossal profits.

The laws of the market work in such a way that only solvent de-mand is fulfilled; that is, only buyers who can pay will have their needs met. Obeying these laws requires willfully ignoring the fact that food is a human right, a right for all.

The speculator in food commodities attacks on all fronts and de-vours everything that might yield him some advantage: he gambles especially on land, agricultural inputs, seed, fertilizer, credit, and foodstuffs. But speculation is a hazardous activity. Speculators can realize in a few seconds a gigantic profit or lose colossal sums.

Two examples: Jérôme Kerviel, a young trader with Société Générale, took positions starting in late 2006 on European stock index futures worth nearly 50 billion euros ($73.6 billion at the January 2008 average exchange rate), an amount greater than the bank's total market capitalization. When his fraudulent trading was discovered in January 2008, Kerviel was accused of having lost 4.8 billion euros ($7.07 billion) for the bank. By contrast, in 2009, the GAIA World Agri Fund managed by the Geneva-based GAIA Capital Advisors, one of the fiercest speculators in agri-food equities, realized a net return on investments of 51.85 percent.

The classic definition of speculation was provided in 1939 by the Hungarian-born British economist Nicholas Kaldor. Speculation, Kaldor writes, is

> the purchase (or sale) of goods with a view to re-sale (or re-purchase) at a later date, where the motive behind such action is the expectation of a change in the relevant prices . . . and not a gain accruing through their use, or any kind of transformation effected in them or their transfer between different markets.

The International Food Policy Research Institute (IFPRI) gives an even simpler definition: "Speculation is the assumption of the risk of loss in return for the uncertain possibility of a reward."

What distinguishes the speculator from any other economic actor is that he buys nothing for his own use. The speculator buys a good—a consignment (or lot) of rice, wheat, corn, oil, and so on—in order to resell it later or immediately with the intention, if the price varies, of repurchasing it later. The speculator is not the cause of price rises, but, as a result of his intervention in the market, he accelerates their upward movement.

There are three categories of operators in the stock market: hedge fund operators, who try to protect themselves against risks linked to variations in asset prices (stock market prices, exchange rates); arbitrageurs, whose activity consists in trading securities (or foreign currencies) with the aim of realizing a profit on the differences in interest rates or asset prices; and, last of all, speculators.

The financial instruments par excellence of the speculator in agricultural commodities are the derivatives and futures contracts (also called forward contracts). A word on their origins. According to Olivier Pastré, one of the leading experts in this area,

> the first derivatives markets were created at the beginning of the twentieth century in Chicago, to aid Midwestern farmers to protect themselves against the erratic fluctuations in commodities prices. But this new type of financial product has, since the beginning of the 1990s, been transformed, from the insurance products that they once were into a product of pure speculation. In barely three years, from 2005 to 2008, the proportion of activity by noncommercial actors in the corn markets increased from 17 to 43 percent.

Agricultural products were bought and sold on world markets for a long time without major problems until 2005. So why did everything so radically change in 2005?

First, the market in agricultural products is very specific. Again, according to Pastré,

> this market is a market of surplus and excess. Only a tiny part of agricultural production is exchanged on international markets. Thus, international trade in cereals represents barely more than 10 percent of production of all crops combined (7 percent for rice). A very small change in world production one way or the other can thus have a drastic effect on the market. A second factor unique to the market in agricultural products is that while demand (consumption) is very rigid, supply (production) is very fragmented (and therefore incapable of being organized and of exerting pressure on changes in prices) and subject more than any other market to climatic fluctuations.

These two factors explain the extreme volatility of market
prices, a volatility that speculation only amplifies.

Until recently, most speculators operated in financial markets.
In 2007, these markets imploded: trillions of dollars' worth of assets were destroyed. In the West, but also in Southeast Asia, tens
of millions of men and women lost their jobs. Governments reduced their social expenditures. Hundreds of thousands of small
and medium-size businesses went bankrupt. Anxiety about the
immediate future and social and financial insecurity became a
way of life in Paris, Berlin, Geneva, London, Rome, and many
other places. Some cities, such as Detroit and Rüsselsheim, were
devastated. In the southern hemisphere, tens of millions more
people sank into the torments of undernutrition, the illnesses associated with hunger, and death by starvation.

The stock market predators, on the other hand, were largely
bailed out by their governments. Public funds financed their lavish
bonuses, their Ferraris, their Rolexes, their private helicopters, and
their luxurious homes in Florida, Switzerland's Zermatt, and the
Bahamas. In short, with the Western governments having shown
themselves incapable of imposing any legal limits on speculators,
banditry in the banking sector flourishes today as never before. But
in the aftermath of the implosion of the financial markets, which
the markets themselves caused, the most dangerous of the tiger
sharks, above all the American hedge fund managers, migrated
to markets in raw materials, especially to the agri-food markets.

The areas in which speculators can operate are almost unlimited. All the goods produced on the planet may become the objects
of speculative bets on the future. In this chapter, I will concentrate
on one kind of speculation, that which affects the prices of food,
especially staple foods, and the prices of arable land.

What are called staple foods—rice, corn, and wheat—account for
75 percent of total world food consumption (rice alone accounts for
50 percent). Twice in recent years, in 2008 and 2011, speculators
have caused a sudden spike in food prices. The spike in staple food

prices in 2008 sparked, as I have noted above, the famous "hunger riots" that shook thirty-seven countries. Two governments were overthrown by the impact of the riots, in Haiti and Madagascar. The images of women in the Haitian shantytown of Cité Soleil baking mud cakes for their children to eat were broadcast in perpetual rotation on television. For several weeks, urban violence, raids of stockpiled food, and protests that brought hundreds of thousands of people into the streets of Cairo, Dakar, Bombay, Port-au-Prince, and Tunis demanding enough bread to survive made the front pages of newspapers everywhere. The world suddenly realized that in the twenty-first century, people were dying of hunger by the tens of millions. Then everything went back to normal. Public interest in the millions of starving people fell back to its normal level and indifference reigned once more.

Many factors are at the source of the rise in prices of staple food products in 2008: the increase in global demand for biofuels; drought, and the resulting poor harvests in certain regions; the lowest level of world stockpiles in cereals in thirty years; the increased demand in emerging countries for meat and therefore grain; the elevated price of oil; and, above all, speculation.

Let us consider the crisis of 2008 in more detail. The market in agricultural products reflects the equilibrium between supply and demand, and is affected by the rhythms of whatever affects these forces, such as climatic fluctuations, which constantly alter this equilibrium. This is why a minor incident in one corner of the planet, because of its eventual repercussions on the global volume of food production (decreasing supply), at the same time as the world's population continues to grow (increasing demand), may have considerable repercussions on the markets and cause spike in prices.

The crisis of 2008 is thought by some to have been unleashed by El Niño, beginning in 2006. Whether or not this is true, when we consider the fluctuations in the global prices of grains in the following graph, we can see clearly that prices began to rise steadily in 2006 and then shot up in 2008, spiking to extremely high peaks. In 2008, the FAO's price index reached an average 24 percent higher than in 2007, and 57 percent higher than in 2006.

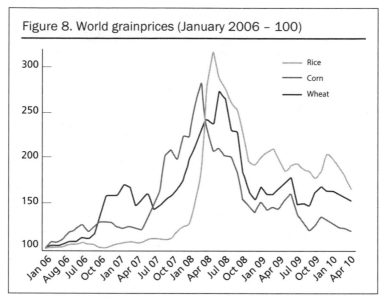

Figure 8. World grainprices (January 2006 – 100)

As Philippe Chalmin explains,

> in March [2008] in Chicago, wheat of standard grade approached $500 per ton. In Minneapolis, a superior grade, Dark Northern Spring Wheat, even reached $800. In the Mediterranean, hard wheat, the type from which pasta and couscous are made, cost more than $1,000. . . . But the crisis was not limited to wheat. The other most important subsistence cereal, rice, followed nearly the same price curve, seeing prices rise in Bangkok from $250 to $1000 per ton.

As for corn, the American bioethanol industry and the approximately $6 billion in annual subsidies to the producers of green gold considerably reduced U.S. supply to the world market. Moreover, since corn is an important contributor to livestock feed, its scarcity on the markets, while demand for meat was (and still is) increasing, also contributed to the rise in prices in 2006. Under normal circumstances, the global cereals harvest reaches about 2 billion tons, of which about one-quarter goes to feed livestock. An increase in demand for meat thus necessarily causes a substantial reduction in the amount of cereal available on the market. Then, in 2008, floods struck the American corn belt, the breadbasket of the American Midwest, especially in Iowa, further increasing the price of corn.

Chalmin shows the twin dimensions—economic and moral—of the activity of speculators operating in the markets for agricultural commodities: "That speculation on the prices of wheat is allowed may seem shocking, even immoral, and it reminds us of an entire history of monopolization and manipulation of prices to the profit of a few dubious financiers." But for the speculators, agricultural products are products on the market like any others. They show no particular consideration for the consequences that their activity may have on millions of human beings as a result of increased prices. They're just "going bull," betting that prices will rise, that's all.

In the event, the tiger sharks were a little late in smelling blood. But as soon as they had spotted their prey, they attacked vigorously. Laetitia Clavreul describes what happened:

> The hedge funds rushed into the agricultural markets, causing an increase in volatility. . . . Agricultural commodities have become commonplace as objects of market activity. Starting in 2004, the hedge funds began to get interested in this sector, which they judged to be undervalued; this explains the development of the futures markets. In Paris, the number of wheat contracts rose from 210,000 to 970,000 between 2005 and 2007.

Speculation in food products reached such proportions that even the U.S. Senate was worried. The Senate denounced "excessive speculation" in the wheat markets, criticizing especially the fact that some commodity index traders were holding up to 53,000 contracts at one time. The Senate also criticized the fact that "six commodity index traders are currently authorized to hold a total of up to 130,000 wheat contracts at a time, instead of up to 39,000 contracts, or one-third less if standard position limits were applied."

Confronted with the insane surge in prices, the major exporting countries closed their borders. Fearing undernutrition and hunger riots in their own territory, they suspended their exports, further aggravating scarcity in the markets and amplifying the rise in prices. As Clavreul writes, "Many producer countries . . . blocked or limited their exports, first of wheat (Ukraine, Argentina, and so on), then rice (Vietnam, India, and so on)."

One day in May 2009 in Senegal, accompanied by Adama Faye, an agronomist and overseas development adviser to the Swiss embassy, and his driver, Ibrahima Sar, I am en route with my wife, Erica, heading north toward Senegal's big plantations. I have brought with me—and have spread out on my knees—the most recent statistical tables from the African Development Bank. The asphalt road is straight, monotonous. Baobab trees line the way; the earth is yellow and dusty despite the morning hour. It is stifling inside the old black Peugeot.

I question Faye uninterruptedly. He is a placid man, full of good humor, extraordinarily good at what he does. But I can sense his rising agitation; my incessant questioning seems to annoy him.

We are crossing the Ferlo. There are hardly any young people left in this semiarid pastoral region. The Ferlo was once home to half a million inhabitants. Tens of thousands of them have migrated to the shantytowns of Dakar. Others have taken the risk of a nighttime ocean crossing to the Canary Islands. Some have disappeared with all their belongings, and their bodies have never been found.

There is not enough water. Rail service from Dakar to Saint-Louis, 320 kilometers (200 miles) away in the northwest, ended long ago. The rails rust peacefully in the sun and are covered over by sand. Soil erosion, the government's negligence, and the deep poverty that leaves people stranded in lethargy have prevailed over the life energy of this magnificent country.

We arrive in the cattle market town of Louga. We are still 100 kilometers (62 miles) from Saint-Louis. Suddenly Faye stops the car. "Come! Let's go see my little sister. . . . She doesn't need your statistics to tell you what's going on."

A meager market, just a few stalls by the side of the road. Mounds of *niébé* (cowpeas) and manioc, a few chickens cackling inside their cages. Peanuts, a few wrinkled tomatoes, potatoes. Oranges and clementines from Spain. Not a sole mango, though Senegal is very famous for them. Behind one of the wooden stalls, a cheerful young woman in a bright yellow *kaftan* and matching head scarf sits chatting with her neighbors. Faye introduces us to Aisha, who is in fact his cousin. She is keen to answer my questions, but the more she talks, the more I sense her mounting anger. Her neighbors in the adjacent stalls join in. Soon a joyous, noisy crowd of children of all ages, young people, and old women has gathered around us on the dusty verge of the road to the north. Everyone wants to have their say, to express their indignation.

A price of a 50-kilogram (110-pound) sack of imported rice has gone up to 14,000 West African CFA francs (about $35). As a result, soup for the evening meal has become more and more watery, with only the few grains a housewife can spare floating in the pot. No one in the nearby area is any longer able to buy more than a quarter sack of rice—to buy a whole sack is out of the question. At the grocer's, the women are now buying rice by the cupful. In the last few years a small bottle of gas has gone up from 1,300 to 1,600 CFA francs, a kilo of carrots from 175 to 245, and a baguette from 140 to 175, while a tray of thirty eggs has risen in one year from 1,600 to 2,500. For fish, same story. The men who, in their small white trucks, bring dried fish from the city of M'bour on the Petite Côte, south of Dakar, are asking 300 francs CFA per kilogram.

Aisha is angry, speaking in a loud voice, sometimes laughing, her laughter bright, joyful, like spring rain. Now she makes a show of scolding her neighbors, who are too timid, in her opinion, in their descriptions of their situation. "Tell the *toubab* [white man] what you pay for a kilo of rice! Tell him! Don't be afraid. The prices for everything are going up almost every day."

"Whose fault is that?" I ask.

Aisha replies, "The truckers—they're thieves. . . ." All goods arrive by road, because the government has shut down the railway.

Faye interrupts to defend the truckers: "At the pump, a liter of gas costs 618 CFA francs; a liter of diesel, 419 francs."

Aisha has pointed out a serious problem often neglected by statisticians, who focus only on the prices of imported food, but not costs added by such factors as transportation. Rice is the staple food in Senegal. The government imports about 75 percent of the rice consumed in the country annually, dealing with the multinational corporations that dominate the market. This rice is sold FOB (free on board) to the Senegalese government. In other words, the price does not include the cost of insurance or transport. Now, in 2008, the price of oil spiked to $150 a barrel on the spot market in Rotterdam. Aisha and her seven children paid the bill. In Louga, and throughout the Ferlo, the prices of essential goods practically doubled in one year. Of course, oil is also prey for the tiger sharks. This is how high finance slowly devours the economy.

Let us consider next the first months of 2011. The most recent explosion in prices has a bitter taste of déjà vu. The World Bank has written:

> The World Bank's food price index rose by 15% between October 2010 and January 2011, is 29% above its level a year earlier, and only 3% below its June 2008 peak. . . . The increase over the last quarter is driven largely by increases in the price of sugar (20%), fats and oils (22%), wheat (20%), and maize (12%).

The World Bank estimates that at least 44 million men, women, and children belonging to vulnerable groups in low- and middle-income countries have, since the beginning of 2011, entered the shadow world of the undernourished, whose lives are scarred by hunger, family disintegration, extreme poverty, and fear of tomorrow. As the World Bank reports:

> Higher global wheat prices have fed into sharp increases in domestic wheat prices in many countries. The transmission rate of global wheat price increases to the domestic price of

wheat-related products has been high in many countries. For instance, between June 2010 and December 2010, the price of wheat increased by large amounts in Kyrgyzstan (54%), Bangladesh (45%), Tajikistan (37%), Mongolia (33%), Sri Lanka (31%), Azerbaijan (24%), Afghanistan (19%), Sudan (16%), and Pakistan (16%).

The report continues:

Maize prices have increased sharply and are affected by complex linkages with other markets. In January 2011, maize prices were about 73% higher than [in] June 2010. These increases are due to a series of downward revisions of crop forecasts, low stocks (U.S. stocks-to-use ratio for 2010/11 is projected to be 5%, the lowest since 1995), the positive relationship between maize and wheat prices, and the use of corn for biofuels. Ethanol production demand for corn increases as oil prices go up, with sugar-based ethanol less competitive at current sugar prices.

Furthermore:

Domestic rice prices have risen sharply in some countries and remained steady in others. The domestic price of rice was significantly higher in Vietnam (46%) and Burundi (41%) between June-December 2010. Indonesia (19%), Bangladesh (19%), and Pakistan (19%) have increased in line with global prices. These Asian countries are large rice consumers, especially among the poor.

Practically all the experts (except, naturally, the speculators themselves) recognize that the evidence shows how speculation plays a determining—and harmful—role in sudden increases in food prices. Two important commentators should be cited here.

First, Olivier de Schutter, my successor as UN Special Rapporteur on the Right to Food:

There would not have been any food crisis without speculation. Speculation was not the only cause of the crisis, but it
accelerated and aggravated the crisis. Agricultural markets
are naturally unstable, but speculation amplifies the brutal
increases in prices. . . . Speculation makes it difficult to plan
production and can brutally increase the cost of food for
countries that import food products.

Heiner Flassbeck was secretary of state in the finance ministry under Minister Oskar Lafontaine in the first administration
of Chancellor Gerhard Schröder from 1998 to 1999. Today he
is Chief of Macroeconomics and Development at UNCTAD in
Geneva, and also the acting director of UNCTAD's Division on
Globalization and Development Strategies. He is one of the most
influential economists on the planet. Together with more than a
hundred collaborators, he directs the most important research
unit in the entire UN system. According to a report released by
UNCTAD, prepared by Flassbeck's team:

The impact of the sub-prime crisis has spread well beyond
the United States, causing a widespread squeeze in liquidity
and credit. And price hikes in primary commodities, fuelled
partly by speculation that has shifted from financial instruments to commodity markets, adds to the challenge for poli-
cymakers intent on avoiding a recession while at the same
time keeping inflation under control.

In January 2011, at its annual meeting in Davos, in the Swiss canton of Les Grisons, the World Economic Forum (WEF) classified
the rise in primary commodity prices, especially food commodities, as one of the five greatest threats to well-being confronting
the world's nations, together with cyberwarfare and the possession by terrorists of weapons of mass destruction.

Klaus Schwab, founder of the Forum, has an astute—and
profitable—system for deciding who is allowed admission to it.

The Forum is essentially a club of the world's one thousand richest corporations. As the WEF itself says,

> Our members represent the 1,000 leading companies and 200 smaller businesses—many from the developing world. . . . Our members are influential, talented and powerful people. . . .
>
> In addition to supporting the World Economic Forum's mission, a typical Member company is one of the world's foremost 1,000 enterprises with a leading role in shaping the future of its industry or region, a solid projected growth potential and a turnover of a minimum of US$5 billion.

Members of this "club of a thousand" pay membership fees of 42,500 Swiss francs (about $46,000), which gives them access to the WEF's meetings. Among them, the tiger sharks are, obviously, numerous. Does the hypocrisy of the world leaders who meet annually in Davos have no limits?

Nevertheless, the opening addresses at the Forum's annual meeting in 2011 in the bunker-like Congress Center in Davos clearly outlined the problem of speculation in agricultural commodities. They even condemned, in the strongest possible terms, the "irresponsible speculators" who, motivated purely by the lure of profit, destroy the markets in foodstuffs and aggravate world hunger. There then followed, over the course of the six following days, a string of seminars, conferences, cocktail receptions, and meetings both public and private in the big hotels of the little snow-covered town, to discuss the issue.

But in the restaurants, the bars, and the bistros that specialize in raclette, the tiger sharks refined their strategies, coordinated their activities, and prepared the next attack on this or that food commodity—or on oil, or some national currency. It is not at Davos that the problem of world hunger will find its solution.

Chalmin asks, "What kind of civilization is this that has found no better way than gambling—the speculative anticipation of profit—to determine the price of bread, or a bowl of rice, for people to eat?" Market logic and the right to food are in absolute contradiction with each other. Speculators play with the lives of millions of human beings. Reason alone demands the immediate and total abolition of speculation in food products.

To vanquish all the tiger sharks once and for all, to protect markets in agricultural commodities from their repeated attacks, Flassbeck himself also supports a radical solution: "We have to snatch primary commodities, especially in food, away from the speculators," he writes. The German verb Flassbeck uses, *entreissen* (to snatch or wrench away from), shows that he is perfectly well aware of the tough fight that awaits those who intend to lead the way.

Flassbeck is calling for a specific mandate from the UN, which, he contends, should put UNCTAD in control of setting prices for agricultural commodities on commodities exchanges worldwide. Going forward, under Flassbeck's system, only producers, traders, and users of agricultural commodities would be permitted to engage in trading on the futures markets. Anyone trading a consignment of wheat, rice, oil, and so on would be obliged to deliver the goods traded. It would be further advisable to impose on traders a high minimum level of self-finance. Anyone who would not make use of a given traded good would be effectively excluded from the commodities exchanges.

If it were implemented, the "Flassbeck method" would keep the tiger sharks away from the basic means of survival for the world's poor, and radically hinder the financialization of agri-food markets. The proposal by Flassbeck and UNCTAD is vigorously supported by a coalition of research organizations and NGOs. Their reasoning is summarized in a remarkable report entitled *When Speculation Matters* by Miguel Robles, Maxime Torero, and Joachim von Braun of IFPRI, who were, respectively, a postdoctoral fellow at IFPRI, the director of its Markets, Trade, and Institutions Division, and its director general from 2002 to 2009.

To oppose this project by claiming that ending speculation on the agri-food markets would amount to jeopardizing the free market is obviously absurd. But what is lacking for now is the will of national governments to implement it.

27

GENEVA, WORLD CAPITAL OF
AGRI-FOOD SPECULATORS

M arc Roche is merely stating the obvious when he writes that the battle against speculation "is equally inseparable from the struggle against the tax havens where the hedge funds are domiciled. And yet, to this day, the G8/G20 countries reveal their sheer hypocrisy, covertly protecting the very thing they publicly condemn. . . . Efforts at regulation also collide with the bank lobby."

"Offshore" capital or other financial holdings are those that are held outside their country of origin; 27 percent of all offshore holdings are held in Switzerland. Fiscal legislation varies from canton to canton in the Swiss confederation. In Zoug, for example, holding companies pay only 0.02 percent in taxes; 200,000 are registered in Zoug. Under Swiss law, anyone who has been a resident of Switzerland for at least ten years and has no desire to seek employment in the country (because he or she is retired or independently wealthy) is eligible for lump sum taxation (*forfait fiscal* in French). The tax is calculated—somewhat vaguely, to say the least—on the basis of the taxpayer's and his or her family's expenditures. In the cantons of Geneva, Vaud, and Valais, idle rich foreigners can negotiate the amount of tax they are required to pay directly with the cantonal government. Despite a few concessions obtained by the European Union and the OECD, banking secrecy remains the supreme law of the land. The Swiss franc is now the second-most-important reserve currency in the world, behind the euro but ahead of the dollar.

The banking lobby is all-powerful in Geneva. This marvelous little republic, which lies on the Rhône River at the southern end of Lake Geneva, covers 247 square kilometers (95.4 square miles) and has a population of just under 465,000. Yet it is the sixth-most-important financial center on the planet. Geneva is also a fiscal paradise that shelters the assets of powerful figures from all five continents. And since 2007, Geneva has become the world capital of speculation as well, especially in agricultural commodities. In this sector, Geneva has now surpassed the City of London. Many hedge funds, whose investment and trading activities rely on anticipating movements in markets—that is, on speculation—have moved to Geneva. One example: Jabre Capital Partners, managed by the Lebanese-born Philippe Jabre, who manages $5.5 billion. Attracted by the extreme leniency in fiscal matters of its current minister of finance, the ecologist David Hiler, traders in agricultural commodities have flocked to the Republic and Canton of Geneva.

Geneva's banks, logically, finance speculators, making available to them the lines of credit indispensable to the transportation, from one end of the planet to the other, of colossal cargoes of rice, wheat, corn, oil-plant products, and so on. The most powerful inspection, verification, testing, and certification services company in the world, SGS S.A. (formerly the Société Générale de Surveillance), which employs more than ten thousand people just to inspect the world's principal port facilities, is, moreover, headquartered in Geneva. The volume of business conducted in Geneva involving primary commodities, a great part of which involves agricultural commodities, reached $1.5 billion in 2000, $12 billion in 2009, and $17 billion in 2010. In addition, in 2010, the Swiss National Bank estimated the total amount of deposits in exchange-traded investment funds in Switzerland at 4.5 trillion Swiss francs (more than $4.85 trillion), a sum equivalent to five times the Confederation's entire budget. But only one-third of this astronomical sum is held in Swiss investment funds, that is, in funds whose management falls under Swiss law. The majority of the hedge funds operating in Switzerland are registered in the Bahamas, the Cayman Islands, Curaçao, Jersey, Aruba, Barbados, and so on, and therefore entirely evade any legal control in Switzerland.

Practically all the Western governments have legislation that restricts the activities of the equity and investment funds registered within their territory. Yet offshore hedge funds are subject to none of these restrictions because the places where they are registered, by definition, have no legislation governing their activities. This is precisely what makes these hedge funds so attractive. They act, to be sure, via a Swiss bank account; in banking jargon, they are "domiciled" in a bank in Geneva. But they are not, I repeat, *not* registered in Switzerland.

Hedge funds constitute the speculative instrument par excellence. They enable the most flagrantly risky trades—and the most profitable. They practice short selling, the sale of financial instruments that they do not own, and typically rely on leverage—that is, debt—borrowing capital for their own use collateralized with the capital they receive from their investors.

In the jungle of Genevan high finance, competition is tough. For the hedge funds and other agri-food funds, the company presentation to potential investors is critical, involving video, statistical reports, graphs, and so on, through which each speculative fund attempts to attract and seduce clients. The name and the symbols of Geneva, the city of John Calvin—the fountains, the view of Mont Blanc, the cathedral, the Reformation Wall—figure prominently in these presentations: the most important thing is to reassure the potential client, to suggest—and why not?—that the hedge fund in question (registered in the Caymans or Curaçao, and so on) is subject to Swiss law. The political stability of the Republic and Canton of Geneva, the honesty of most of its citizens, the solidity of its institutions, the stony stolidity of its bankers are all convincing arguments aimed at investors, wherever they come from—France, the United States, Qatar, or Australia.

But the reality is something else entirely. The vast majority of hedge funds, as I have said, are not covered by Swiss law. Nor are they subject to the control of the Swiss Finanzmarktaufsicht (FINMA), the country's Financial Markets Authority. The current president of FINMA (as of January 2011), Anne Héritier Lachat, admits, "We don't regulate offshore hedge funds because the law does not give us the jurisdiction to do so."

There is thus no control over two-thirds of the speculators roaming Geneva's jungle of high finance. And this fills honest savers and investors with despair. One investor, who lost large sums in the jungle of the Genevan banking system after investing with hedge funds speculating in rice, corn, wheat, and so on, complained, "How can it be . . . that financial companies are permitted to work this way, boasting that they are authorized by FINMA, deceiving us, while in reality they escape any oversight?"

The government of the Republic and Canton of Geneva, however, pampers the tiger sharks. Beyond the multiple tax privileges that it extends to them, Geneva underwrites and sponsors the annual conference that they hold in the city. Hedge fund managers who operate in the agri-food sector met on June 10, 2010, for the JetFin Agro 2010 Conference at the Grand Hotel Kempinski on the Quai du Mont-Blanc. (Founded in 2002, JetFin is a Geneva-based consulting and conference-organization company catering to the "alternative investment" industry.) JetFin held its subsequent Agro conference on June 7, 2011, in the same venue. "Agriculture today is the shining light in the investors' universe," JetFin writes in its 2011 conference brochure, promising that at Agro 2011, top managers will explain how to "realize higher profits in exciting markets." The conference invitation (and its online list of sponsors) is graced by the red-and-gold coat of arms of the Republic and Canton of Geneva, JefFin's "Institutional Partner."

Once again, the government blesses—and funds—the luxurious tank where tiger sharks from all over the world come to swim. The attitude of the Genevan authorities is almost scandalous. To use taxpayers' money and the prestige of Geneva to pamper a few hundred speculators, including the most evil ones, is shameful. Two powerful NGOs, one Catholic, Fastenopfer (the Swiss Catholic Lenten Fund), the other Protestant, Bread for All, in fact sent a vigorous letter of protest on June 28, 2010, to the Geneva government.

Their lordships did not deign to reply.

28

LAND GRABS AND THE RESISTANCE
OF THE DAMNED

Immediately after the food crises of 2008, many of the countries that are rich in capital but poor in land, such as the Gulf states, or that have a high population density, such as China and India, began to buy or rent land on a large scale in other countries in order to ensure their food supply (in cereals or meat), seeking to make themselves less dependent on market fluctuations and to respond to growing demand at home. With the dawn of a new food crisis in 2011, reports of land grabs increased. This phenomenon, added to the growth in purchases of land for the purposes of speculation, confirms that land has become a safe investment, a refuge asset, often more profitable than gold. And in fact, with the price of land on average thirty times lower in developing countries than in the countries of the North, land is an investment that really pays off. Moreover, because the international community has decided not to protect the rights of local populations anytime soon, buying up land for speculative purposes has bright days ahead.

In Africa, in 2010, 41 million hectares (141.3 million acres) of arable land were bought, rented, or acquired without compensation by American hedge funds, European banks, and the sovereign wealth funds of Saudi Arabia, South Korea, Singapore, China, and other countries. The example of South Sudan is particularly instructive. After a war of liberation that lasted twenty-six years and left more than a million dead and wounded, the new state of South Sudan was born on July 9, 2011. But even before the republic's founding, the provisional government in Juba, its capital city, had sold off 600,000 hectares (1.48 million acres) of arable land to the Texas-based Nile Trading and Development, Inc.—1 percent of the nation's territory—for the bargain-basement price of $25,000, or about 2 cents per acre. Nile Trading also has an option on a further 400,000 hectares (988,000 acres).

Speculation can also be "internal." In Nigeria, the rich merchants in Sokoto and Kano have in various ways laid hands on tens of thousands of hectares of land use for food crops, most often by bribing public officials. Similarly dubious transactions are common in Mali. Rich businessmen in Bamako—or, most often, in the Malian diaspora in Europe, North America, and the Persian Gulf—are buying up land. They do not farm the fields they acquire, but wait for the price of land to rise before selling out to some Saudi prince or a New York hedge fund.

The speculators who swoop down on land used for growing food crops, in order to sell the land later or to replant the fields immediately with crops for export, use a wide array of methods to dispossess African farmers of their means of existence.

Bread for All and Fastenopfer have published an in-depth investigative report on the tiger sharks operating in the financial industry of Geneva and Zurich. As the report says:

> In Switzerland, it is above all banks and investment funds that are involved in land grabs. Thus, Crédit Suisse and UBS participated in 2009 in issuing shares on behalf of Golden Agri-Resources. . . . This Indonesian company is grabbing large areas of tropical forest in order to plant vast areas of oil-palm monoculture—with grave consequences for the climate and the local population. Furthermore, Golden Agri-Resources is included in the funds that the two big banks offer to their clientele.

The same chapter in the report continues:

The Sarasin and Pictet funds invest in Cosan, among whose activities is the purchase of land and entire farms in Brazil with the aim of profiting from pushing up land prices. Cosan has been strongly criticized for working conditions approaching slavery on its sugarcane plantations. Many Swiss funds, both conventional and speculative (hedge funds), invest in agriculture, such as GlobalAbriCap in Zurich, GAIA World Agri Fund in Geneva, and Man Investments (CH) AG in Pfäffikon. All of these funds invest in businesses that buy land in Africa, Kazakhstan, Brazil, or Russia.

As the report's introduction concludes:

> In the context of the food crisis, many developing countries have seen themselves criticized for having neglected the agricultural sector. It would seem that some of them are convinced that renting and selling their land constitutes a solution to promoting agriculture, regardless of the grave social, economic, and ecological consequences.

The seizure of land by speculators causes the same social consequences as the acquisition of land by the vultures of green gold. Whether they involve the Libyans in Mali, the Chinese in Ethiopia, or the Saudis and the French in Senegal, such land grabs are clearly detrimental to local populations, and are often accomplished without the people even being consulted beforehand. Entire families are deprived of access to natural resources and driven off their land. When the multinationals do not install their own contingents of workers on-site, a small part of the local population will be able to find work, but for starvation wages and in working conditions that are often inhumane. Most of the time, families are evicted from their ancestral lands; their kitchen gardens and their orchards are soon destroyed, while promises of fair compensation evaporate. And with the eviction of small farmers, it is the food security of thousands of people that is put in danger.

There is also an ancestral know-how, transmitted from generation to generation, that disappears: the understanding of the soil, the slow process of selecting the right seed to suit each plot of land and the amount of sunshine and rain that it receives—all this is swept away in a matter of a few days. In its place, the agrifood corporations plant monocultures of hybridized or genetically modified plants, cultivated according to the systems of industrialized agriculture. The companies enclose the parcels of land in such a way that farmers and nomadic pastoralists no longer even have access to local river shoreline, forest, and pasturage.

Speculating on food products, speculating on land, the traders are in fact speculating on death.

The big multinational French companies active in Africa, such as Bolloré, Vilgrain, and others, boast about the benefits they bring to local populations by investing in their land: construction of infrastructure (roads, irrigation systems, and so on), employment opportunities, increases in national production, knowledge and technology transfer, and so on. In the words of Alexandre Vilgrain, president of CIAN (Conseil Français des Investisseurs en Afrique; French Council of Investors in Africa),

> We might consider that the countries of the South judge the countries of the North, and France in particular, much less on their policies of development aid than on the policies of their businesses that invest locally. . . .
>
> The African continent, where our businesses have a long and deep experience, and for the most part, a common language, is becoming a playing field for global investors.
>
> Our country, and therefore our businesses, have every chance of success there, on the condition that we make more of a team effort.

Vilgrain's "playing field" is Africa's field of desolation. The pillaging of the continent is accompanied by an impressive media blitz—the pillagers enjoy "communicating." Yet in order to cover up their misdeeds, they sometimes invent phrases that are right on target. Among those that get the most use is the famous "win-win." Setting up a win-win relationship, based on satisfying the needs of both parties, makes it possible to resolve conflicts. A win-win agreement is one that enables each party to maximize its interests, and increases the each partner's profits. In short, in losing their land, farmers ensure themselves of many advantages, just like the agri-food corporations that steal the land from them! Thus speculation, so to speak, contributes to the common good and general happiness.

The World Social Forum held in Dakar in February 2011 confirmed that Africa has a civil society of extraordinary vitality. From one end of the continent to the other, people are organizing the resistance to the tiger sharks. Consider the following examples.

SOSUCAM, the Société Sucrière du Cameroun (Cameroon Sugar Company), which belongs to Vilgrain, owns thousands of hectares of land in Cameroon, which is, with Sierra Leone, one of the most corrupt countries on the continent. According to CODEN (Comité de Développement de la Région de N'do; the Committee for the Development of the N'do Region), a Cameroonian coalition of farmers' unions, churches, and other civil society organizations, SOSUCAM signed a ninety-nine-year lease in 1965 with the national government to develop 10,058 hectares (24,854 acres). (SOSUCAM agreed to pay indemnities to the dispossessed farmers, but never did.) In 2006, a second lease added 11,980 hectares (29,603 acres) to the company's holdings. This time, SOSUCAM agreed to pay an annual indemnity to the affected communities, but in the amount of only 2,062,985 CFA francs ($3,944.50 at the average 2006 exchange rate)—the equivalent of about $6.25 per family annually. About six thousand people live on the cropland acquired by SOSUCAM. It is of course useless to point out that they were given no chance to have a say regarding the two transactions, which were concluded between Vilgrain and officials in the capital city, Yaoundé.

According to the resistors, SOSUCAM's increase in its land holdings is at the expense of rural communities that are seeing their food security threatened. Furthermore, working conditions in the plantations are disgraceful and dangerous.

"First they took our land without asking us. The parcels that remain to us are not sufficient to feed our families and our harvests have been altered by the insane level of herbicide use on the plantations. SOSUCAM has even forbidden us to keep livestock . . . ," says Michel Essindi, a farmer and member of CODEN. . . .

The most fertile land, formerly used for growing food crops, is now off-limits. Pollution of the air, soil, and water caused by sugarcane processing, makes our crops rot in the ground, reducing production for local markets. . . . The consequences of the company's activities are thus catastrophic and it seems to have total control over the region. . . .

Only 4 percent of SOSUCAM's employees are former farmers who have lost their land. As workers on the plantations, they do not earn enough to meet their own needs and those of their families. . . . Furthermore, both plantation workers and villagers suffer from the environmental consequences of sugar production and processing. . . .

The parent company of SOSUCAM is the SOMDIAA Group, which has been run by the Vilgrain family since 1947. The Vilgrain family also runs the Grands Moulins de Paris, a leading European milling company and the springboard for their agro-industrial venture in Africa. SOMDIAA notably owns three flour mills, in Cameroon, Gabon, and Réunion; eight sugar plants in various parts of Africa; and tens of thousands of hectares of land in many countries. On SOMDIAA's website, you could until recently find the following edifying sentiment: "human values are the very foundation of the Group."

The mobilization of farmers, union members, religious communities, and urban community organizers united in CODEN has succeeded in preventing Vilgrain and the Cameroonian government from signing a third contract that would have entailed a new round of land plunder and another forced exodus of farming families.

Another example: Benin.

The majority of Benin's 8 million citizens are farmers working small or medium-size holdings, parcels of one or two hectares (2.5 to 5 acres). A third of Benin's people live in extreme poverty, that is, on an income of $1.25 per day or less. Undernutrition affects more than 20 percent of the population.

In Benin, it was originally native Beninese large landowners with close ties to the government who engaged in land grabbing. Threatened with starving to death, farmers sold their lands, often for laughable prices—for a mouthful of manioc. The barons of Benin today continue to operate in the same fashion: "the current prices [for land] are relatively low and the money that the farmers get in exchange for their land is quickly spent. But their means of existence and their source of food is lost forever." The barons accumulate land but let it lie fallow, waiting for prices to rise before selling it again. In short, just as on the real estate market in any European city, the speculators buy, sell, then buy again, then sell again, always exchanging the same properties in anticipation of ever-higher profits.

The Zou region was formerly the breadbasket of Benin, its main wheat-producing area. Today, Zou has the highest level in the country of seriously undernourished children under five. Instead of investing in subsistence agriculture—or in other words, instead of supporting farmers in securing supplies of fertilizers, water, seed, draft animals or farm machinery, and tools, and developing the country's road infrastructure—the national government prefers to import rice from Asia and wheat from Nigeria, which further undermines local farmers.

A former banker close to the foreign "investors," especially the French, Boni Yayi (also known as Thomas Yayi Boni) was elected president of the republic in 2006. On March 13, 2011, he was reelected. On the evening of his victory, Yayi's spokesman warmly thanked the French PR agency Euro RSCG for its "precious support." Euro RSCG is an affiliate of the Bolloré group. In 2009, Bolloré received from Boni a concession for the port of Cotonou. In 2011, throughout the country's seventy-seven communes, Euro RSCG organized the banker-president's reelection campaign, to the tune of millions of euros. The previous year, "foreign donors" (including Bolloré) had financed the development of Benin's LEPI (Liste Électorale Permanente Informatisée; Computerized Permanent Electoral List), which cost 28 million euros (about $37 million in 2010). The opposition had vigorously criticized the LEPI. At least 200,000 potential voters, they said, had been excluded, notably in the southern part of the country, where opposition to Yayi was apparently the strongest. However, Yayi was reelected by a margin of 500,000 votes (with 3.6 million registered voters).

Nestor Mahinou, a leading figure in SYNPA (Association Synergie Paysanne; Farmers' Synergy Association) sums up the disastrous situation of Beninese farmers: "While local small farmers are forced to sell their land because they don't have the means to cultivate it, the vast areas of fertile land bought up by third parties lie fallow." SYNPA, founded in 2000 in Cotonou, is Benin's most powerful movement for the defense of farmers who have had their lands confiscated. With support from ROPPA and its president, Mamadou Cissokho, SYNPA leads an exemplary struggle against the neocolonial system in Benin.

Some sovereign wealth funds, belonging to Asian and African countries, among others, behave no more honestly than private speculators. The example of the Libyan-African Portfolio (LAP) is instructive.

In 2008, the fund was "offered" 100,000 hectares (247,105 acres) of irrigable rice-growing land by the Malian government. In order to exploit this opportunity, the LAP incorporated a private company in-country under Malian law named Malibya, which would enjoy the use of its Malian land holdings for a renewable term of fifty years, without paying any identifiable compensation.

In Mali, water is a critical challenge for agriculture; less than 10 percent of the country's arable land is irrigated. Yet Malibya's contract gives the company unlimited use of "the waters of the Niger River during the rainy season" and of the "necessary quantity of water" the rest of the time. An irrigation canal 14 kilometers (8.7 miles) long, already built and watering 25,000 hectares (61,776 acres) now under Libyan control, is already causing serious problems for farmers and nomads in central Mali. The canal is causing farmers' wells and ponds used by nomadic Fula nomads and their livestock to dry up. Between their migrations, the nomads normally grow sorghum in fields that used to have moist soil but which today are dry.

Mamadou Goïta is one of the principal leaders of ROPPA. Together with his colleagues, especially Tiébilé Dramé, Goïta succeeded in 2008 in compelling the Malian government to publish the contract it had signed with the Libyans. Says Goïta, "The Libyans behave as if they were in conquered territory, as if the land were a desert, even though thousands of Malians live here." Drame adds that the "run on agricultural land in Mali [by foreigners] exacerbates conflicts at the same time as the country is having difficulty in feeding its population. . . . For generations, families have grown millet and rice on these lands. . . . What will become of these people? . . . Those who resist are taken in for questioning, and some are imprisoned."

To the farmers' unions that protest evictions without compensation, Abdalilah Youssef, Malibya's director general, replies with exquisite politeness—and incredible bad faith, saying that he understands "the necessity of reorganizing the local population, that is to say, the villages that are going to leave the area." Goïta and his colleagues have no confidence in the "reorganization of the populations" proposed by Youssef. They are demanding the annulment of the contract with the Libyans, pure and simple. So far, in vain.

The instances of resistance on the part of farmers in Cameroon, Benin, and Mali are exemplary. Here is one more.

By constructing the gigantic Diama dam (completed in 1986) on the Senegal River, about 40 kilometers (25 miles) upstream from Saint-Louis, the country of Senegal gained tens of thousands of hectares of arable land. A large part of this new farmland is today monopolized by the Grands Domaines du Sénégal (GDS). In Senegal, any multinational company, foreign investor, and so on may acquire 20,000 hectares (49,421 acres) or more, so long as it enjoys useful relationships in the capital city, Dakar. Leases may be of unlimited duration, with an exemption from taxation for ninety-nine years.

For the members of the farmers' union of the rural community of Ross Béthio who met me and my team during our visit there in 2009, the GDS constitutes an enemy wreathed in mystery. The GDS belongs to investment groups in Spain, France, Morocco, and so on. The company produces, partly in greenhouses, sweet corn, onions, bananas, melons, green beans, tomatoes, green peas, strawberries, and grapes. On average, 98 percent of its produce is exported by ship through the nearby port of Saint-Louis directly to Europe. The GDS benefits from its "vertically integrated" supply chain: the company's plantings lie in the irrigated floodplain along the banks of the Senegal River in the Walo region. Their own ships (or ships that they lease) transport the produce. In Mauritania and in Europe, the GDS maintains facilities for ripening fruit. The financial groups that own the GDS are in many cases the principal stockholders in French supermarket chains.

Walo is dotted with immense greenhouses sheathed in brown plastic equipped with mechanical watering systems. Despite agronomist Adama Faye's connections in the prefecture in Saint-Louis, we do not succeed in gaining access to any of the GDS's facilities, which are protected by armed guards in blue uniforms, metal fencing four meters (13 feet) high, video surveillance cameras, and so forth. We are stopped in front of the entrance to one of the most gigantic GDS facilities, which belongs to a French fruit company, La Fruitière de Marseille. We negotiate by electronic means with a director barricaded in an administrative building whose outlines we can glimpse in the distance. He has a strong Spanish accent. "You do not have authorization for a visit . . . sorry. . . . Yes, not even the UN can do anything about it. . . . The prefect in Saint-Louis? . . . He has no jurisdiction here. . . . You have to apply to our offices in Paris or Marseilles." In short, none of us will get in.

I use a tactic that has worked for me on other occasions: I don't move. I wait for hours in front of the padlocked gate while the guards give us the evil eye. Finally, toward evening, an Audi Quattro drives up, arriving on the paved road that leads to Saint-Louis. A young French technician who seems friendly, and who it turns out has just started working at the GDS, stops in front of the gate. I approach his vehicle. He defends his employer heatedly: "We pay the surveying costs. . . ." Then, a little later: "Many of our fields are on elevated land, at an altitude of twelve or fifteen meters [40–50 feet]. To water them, you need motorized pumps. Senegalese farmers don't have any. . . . We don't pay taxes? That's not true! We employ young people from the villages. The Senegalese government collects taxes on their incomes." End of conversation.

Situated about 50 kilometers (30 miles) from Saint-Louis on the road to Mali, Ross Béthio is home to more than six thousand co-operative farmers. Djibril Diallo, a warmhearted man in his fifties with bright eyes and a receding hairline who wears a brown djellaba, is the executive secretary of the farmers' union. When we meet, he is surrounded by the members of his committee, four men and three women. The farmers of Walo harvest rice twice a year. But the harvests are modest—1 hectare yields only 6 tons of paddy (threshed, unmilled rice)—and the prices paid by the merchants who come from Dakar to take the rice away in their trucks are low. In 2010, an 80-kilogram (176-pound) sack went for 7,500 CFA francs ($15).

Diallo Sall, Diallo's adjunct, is a lively young man, ironic, impatient. Interrupting Diallo's rather lengthy welcoming speech, he exclaims, "Our wives, our young people go out in the rice fields without having eaten first. In the fields, they eat wild fruit. . . . If we tell the health worker about this, he replies, 'You're up against power, you are their opponents.'"

Despite their modest means, the hospitality of the Senegalese is lavish. A table is laid in the hut that serves as the headquarters of the union's leadership committee, near the local mosque. Fans squeak and creak. A delicious aroma escapes from the kitchen. Grilled carp caught in the river, onions, chicken, and potatoes are waiting for us in big metal bowls.

The rice farmers of Ross Béthio are fighters. The savvy way in which they go about their campaign of resistance impresses me. Their union is affiliated with ROPPA and, on a global level, with La Via Campesina. For the farmers, the GDS is beyond reach. But the subprefect and the prefect in Walo and several government ministers in Dakar are targets they could hit.

The alienation of land in Senegal proceeds as follows. Rural land belongs to no one, and therefore lies in the hands of the state. There is no system of land registration. But farming communities posses an unlimited right of usufruct (the right of free use and enjoyment so long as the land is not damaged or destroyed) over the lands they occupy, a right proceeding from immemorial custom. The government has created an institution particularly intended to act in such matters, the rural councils. These are obviously dependent upon the party in power in Dakar. Their jurisdiction is important: they are responsible for boundary marking; that is, they undertake the drawing of boundaries on rural land previously held by the state and used as commons. They assign surveyed and enclosed land to its new proprietors.

The charges drawn up by the Ross Béthio farmers' union are serious but well documented: the confiscation of land for the benefit of the GDS relies upon secret negotiations that take place in Dakar. The rural councils responsible for determining the land's boundaries—that is, for alienating land to the benefit of the GDS—take their orders from the government. The surveyed boundaries are recorded in an official document that must be validated first by the local subprefect, then by the prefect, and finally by a minister. Yet the union contends that some federal officials responsible for validating the boundaries, and even some government ministers in Dakar, have added to the amount of alienated land several thousand hectares intended for their own private use. A rural council determines the boundaries of a given area of arable land and assigns it to a given GDS installation. As the document confirming the transfer makes its way through the bureaucratic jungle, the amount of land stolen from the farmers keeps increasing.

Who profits from the farmers' dispossession? According to the union, it is above all the GDS that profits, of course, but also—and to varying degrees—certain subprefects, certain prefects, certain ministers, and many of their friends. By mobilizing mass opposition to the land grabs, by increasing their public declarations on the international stage, and by initiating legal proceedings in Senegalese courts, Diallo, Sall, and the members of their union, who farm rice, vegetables, and fruit and raise livestock in Walo, are fighting the destruction of their livelihood—with a determination and courage that compel admiration.

29

THE COMPLICITY OF THE WESTERN STATES

The ideologues at the World Bank are infinitely more danger-ous than the PR hacks who work for Bolloré, Vilgrain, and company. With hundreds of millions of dollars in credits and sub-sidies, the World Bank in effect finances the theft of arable land in Africa, Asia, and Latin America.

For Africa, the World Bank's ideologues have developed the following theory to justify this program of theft: On 1 hectare planted with millet, farmers in Benin, Burkina Faso, Niger, Chad, or Mali harvest only 600 to 700 kilograms of grain per year (536 to 625 pounds per acre) in normal weather—and "normal" weather is rare. In Europe, however, as I have said, 1 hectare produces 10 tons of wheat. Therefore, it is more worthwhile to entrust to the tender care of the agri-food conglomerates—with their capital re-sources, their skilled technicians, their marketing and distribution networks—the land from which the poor Africans are incapable of extracting higher yields.

For most of the Western ambassadors sitting on the UN Human Rights Council, the World Bank's word is gospel. I re-member Friday, March 18, 2011, in the main hall of the Human Rights Section, on the first floor of the east wing of the United Nations building in Geneva. Davide Zaru is a young Italian ju-rist with a lively intellect, a confirmed talent for diplomacy, and a total commitment to the right to food. He is currently international relations officer at the Human Rights Unit of the European Commission, Directorate-General for External Relations, in Brussels, which is directed by Catherine Ashton, Baroness Ashton of Upholland, a British Labour politician who in 2009 became the European Union's High Representative of the Union for Foreign Affairs and Security Policy. When the UNHRC is in session, Zaru stays in Geneva. At the UN, it is his job to coordinate the votes of those among the twenty-seven member states of the European Union whose representatives sit on the council. That morning, Zaru had a desperate air. "I can't help you!" he cried. "Explain my situation to our friends in La Via Campesina. In its current form, the resolution will not pass. The Western countries are absolutely opposed to it. They don't want a convention on the protection of the rights of farmers."

Supported by numerous confederations of farmers' unions, NGOs, and governments of countries in the southern hemisphere, the Human Rights Consultative Committee of the UNHRC had, over the course of three years, developed a report on the protection of the rights of small farmers. In its recommendations, the committee demanded that the UN adopt an international convention that would allow dispossessed farmers to defend their rights to their land, seed, water, and so on, against the vultures of green gold and other tiger shark speculators. The resolution was directly inspired by La Via Campesina, and in particular by the Dakar Appeal Against the Land Grab issued at the end of the World Social Forum 2011:

> Considering that recent massive land grabs targeting tens of millions of acres for the benefit of private interests or third states—whether for reasons of food, energy, mining, environment, tourism, speculation or geopolitics—violate human rights by depriving local, indigenous peasants, pastoralists and fisher communities of their livelihoods, by restricting their access to natural resources or by removing their freedom to produce as they wish, and exacerbate the inequalities of women in access and control of land;

Considering that investors and complicit governments threaten the right to food of rural populations, that they condemned them to suffer rampant unemployment and rural exodus, that they exacerbate poverty and conflicts and contribute to the loss of agricultural knowledge and skills and cultural identities;

Considering also that the land and the respect of human rights are firstly under the jurisdiction of national parliaments and governments, and they bear the greatest share of responsibility for these land grabs;

We call on parliaments and national governments to immediately cease all massive land grabs current or future and return the plundered land. . . .

The prospect of seeing this new instrument of international law enforced horrified the Western governments, especially the American, French, German, and British administrations, which are often closely tied to the great predators of the agri-food industry. Such a convention in international law negotiated, signed, and ratified by the world's governments would by its very nature civilize the jungle of the free market a little bit. How dare they!

Worse still, the resolution on the convention included a statement of the rights of all signatory states' citizens to their lands and obligated those states to set up the tribunals necessary to make these rights enforceable. Note that, in this regard, the council would have been creating an innovation in international law. In Senegal, Mali, Guatemala, Bangladesh, and other countries in the southern hemisphere, for a farmer to take to court, for example, one of the vultures of green gold or a speculator based in Paris, China, or Geneva, is sometimes too dangerous, or even simply impossible. Local judges cannot be relied upon to be independent, and the adversary is too powerful. The council therefore recognized states' "extraterritorial responsibility." But, as a result, if France signed and ratified a convention on the protection of the rights of farmers, then the French government would be responsible for the conduct of Bolloré, Vilgrain, and companies like La Fruitière de Marseille in Benin, Senegal, or Cameroon. Dispossessed African farmers and their unions would be able to have recourse to the French justice system. We may better understand why, faced with such dire prospects, the Western governments mobilized all their diplomatic resources to sabotage the resolution initiated by the farmers' unions of the South and endorsed by the Consultative Committee.

The Consultative Committee comprises international experts elected on a proportional basis according to the population of the world's continents. The Human Rights Council, however, is an intergovernmental body comprising representatives of forty-seven countries. For the council to debate the recommendations formulated by the Consultative Committee, one of the member states of the council must propose the resolution. At the sixteenth session of the council, in March 2011, the resolution on the development of a convention protecting the rights of farmers was proposed by Rodolfo Reyes Rodriguez, vice president of the council and the Cuban ambassador to the UN. A brilliant diplomat, Reyes Rodriguez is no softie. After volunteering to fight against the South African expeditionary force during the Angolan Civil War, he lost a leg in 1988 in the decisive battle of Cuito Cuanavale. But the obstructiveness of the Western ambassadors forced him to modify the resolution. For the time being, the fate of the new convention on the protection and justiciability of farmers' rights remains uncertain.

EPILOGUE

But where there is danger,
what saves us also grows.
— Friedrich Hölderlin, "Patmos"

The Earth has a surface area of 510 million square kilometers (197 million square miles). Just under 71 percent of that
total area is water; 7 billion human beings live on the rest. We
are very unequally distributed, with some areas empty and some
overpopulated, as a result of natural features (the glacial polar
regions, deserts, semiarid areas, mountain ranges, valleys and fertile plains, maritime coasts, and so on) and economic realities (agriculture, livestock rearing, fisheries, industry, city, countryside,
and so on).

The first task of the living species that comprise the natural
world, plants, animals, and human beings, is to feed themselves in
order to live. Without food, the organism dies. The second task is
to reproduce. In order to reach maturity and adulthood, the age
at which species can give birth to their offspring, and to be able to
procreate, to give life to a new being, all creatures absolutely must
feed themselves.

It is in order to feed ourselves that we men and women have
gathered, hunted, fabricated weapons and tools, embarked on

migrations and voyages. It is in order to feed ourselves that we have worked the earth, seeding, planting, making still more tools, and that we have sought to gain knowledge of plants and have domesticated animals. It is, again, in order to feed ourselves that we humans have developed, like animals, an obsession with territory, and determined the limits within which we feel at home, and defended this space against those who might covet it. And the desire of others has always been all the sharper when our territory is richer or contains some hidden treasure, when it offers some particular advantage.

Beyond the first stage of agriculture, in the course of which men and women set themselves to fabricate still more tools, storage containers, and clothing, and to improve their habitat, artisanal production developed. It became necessary to exchange goods, to trade, to travel. The economy and its infinite development were born of the necessity of men and women to satisfy our needs, foremost of which is to feed ourselves and our children. Babies cry when by chance they are forgotten and feel hungry. Crying is their only means of expressing themselves; they will shriek for hours, until they can cry no more. When babies exposed to famine lose their strength, they also lose their ability to express their needs by crying, and their voices are stilled.

Today, half of the children born in India are seriously and permanently undernourished. For them, every passing moment is suffering. Millions will die before age ten. Others will continue to suffer in silence, to vegetate, to seek in sleep some relief from the pain that devours their guts.

In the beginning of human history, taking food away from others was something done by strong men, when their women and children had absolute need of it. But the time when the inelastic needs of human beings must face a quantity of goods insufficient to meet those needs is today long past. And if a billion people suffer from hunger, it is not because the world does not produce enough food, but because the powerful hold a monopoly on what the Earth provides. In this finite world of ours, in which there are no new places left to discover, nor any new lands left to conquer, the monopolizing of the Earth's bounty takes on new meaning. It is an immense scandal. The lords of the agri-food markets and the agricultural commodities exchanges decide every day who on this Earth will live or die. They are filled by only one obsession: profit.

In 1947, Mahatma Gandhi remarked, "Earth provides enough to satisfy every man's need but not for every man's greed." Josué de Castro was the first to show that the principal factor responsible for hunger and the lives lost to undernutrition is the unequal distribution of our planet's riches. Yet, since de Castro's death some forty years ago, the rich have grown still richer and the poor infinitely poorer. Not only has the financial, economic, and political power of the global agri-food corporations grown enormously but the individual wealth of the most well-off people has seen exponential growth. Eric Toussaint, Damien Millet, and Daniel Munevar have analyzed the trajectory of billionaires' fortunes in the course of the last ten years. Here are the results of their study. In 2001, there were 497 dollar billionaires in the world, and their total aggregate wealth amounted to $1.5 trillion. Ten years later, by 2010, the number of dollar billionaires had increased to 1,210 and their total wealth to $4.5 trillion. The total wealth of these 1,210 billionaires exceeds the gross national product of Germany.

The collapse of financial markets in 2007–8 destroyed the lives of tens of millions of families in Europe, North America, and Japan. According to the World Bank, 69 million more people were thrown into the abyss of hunger. In the countries of the South, everywhere, new mass graves have been dug. Yet, by 2010, the wealth of the very rich had surpassed the level reached before the collapse of the financial markets less than three years before.

Who are the great powers in the global agri-food order that today control humanity's food? They are first of all the few global private corporations that dominate the markets in question. These monsters of global trade, with their tentacles everywhere, control the production and the selling of the inputs that farmers and stockbreeders must buy—seed, organic and mineral fertilizers, phytosanitary products (which fight pests and pathogens), pesticides, fungicides, and so on. The companies' traders are the principal operators on the world's commodities exchanges. They are the ones who determine the prices of food. Going forward, water will also be in large part under the control of these corporations. And in only a few years, they have acquired tens of millions of hectares of arable land in the southern hemisphere.

But the world's economic order, dominated by global corporations, hedge funds, and big international banks, is not part of the *natural* order. There is nothing "natural" in market forces. It is the market ideologues who, in order to legitimate the murderous practices of global capital and to appease the consciences of financial operators, claim that these "laws of the market" are natural, always referring to them as if they were like the "laws of nature."

A multitude of causes are involved in the chronic undernutrition of one person in six or seven on the planet and in the death from hunger of a scandalous number of people. But, as we have seen throughout this book, whatever the causes are, humanity has at its disposal the means to eliminate them. In his famous Elmhirst Lecture, delivered at the triennial meeting of the International Association of Agricultural Economists in Malaga, Spain, on August 26, 1985, Amartya Sen said, "In the field of hunger and food policy, the need for speed is of course genuinely important." Sen is right: there is not a second to lose. To wait, to bicker over means, to get lost in Byzantine debates and complicated discussions—all the "choral singing" that so shocked Mary Robinson when she was UN High Commissioner for Human Rights—is to become an accomplice to the monopolists, the hoarders, the predators.

The solutions to the problem of world hunger are known; they cover thousands of pages of project proposals and feasibility studies. In September 2000, it was reported that of the UN's then 192 member states, 146 rushed to send their representatives to New York to draw up a list of the principal tragedies afflicting humanity on the threshold of the twenty-first century—hunger, extreme poverty, polluted water, infant mortality, discrimination against women, AIDS, epidemics, and so on—and to decide on objectives, targets to aim for in the battle against these scourges. The heads of state and heads of government calculated that, to put an end to all these tragic afflictions—with hunger at the top of the list—would require an annual investment of $80 billion for fifteen years. And in order to raise that much money, it would be enough to impose a 2 percent tax on the wealth of the 1,210 billionaires that there were in the world in 2010.

But how can we curb the insanity of *les affameurs*, the "starvers," those who exploit famine situations for financial gain? First, by combating the corruption of the leaders of many of the countries in the southern hemisphere, their venality, their appetite for the power that their positions afford them and for the money that those positions are liable to bring. The misappropriation of public funds in some Third World countries, the enrichment of elected officials at public expense, is a calamity. Where corruption is rife, countries are sold to the predators of global finance, who can then just help themselves to the whole world.

Paul Biya, the president of Cameroon for the last thirty years, spends three-quarters of his time at the Intercontinental Hotel in Geneva. Without his active complicity, Alain Vilgrain's conglomerate would not be able to seize tens of thousands of hectares of arable land in central Cameroon. And without Biya's complicity, Vincent Bolloré would not have secured the privatization of the state enterprise Socapalm and would not have been able to seize another 58,000 hectares. When, in Las Pavas, in the department of Bolívar in northern Colombia, the paramilitary killers in the pay of Spanish global palm oil corporations drive farmers from their land, they are "authorized"—that is, encouraged—by the country's leaders. The current president, Juan Manuel Santos, is known to have links to the Spanish companies, just as his predecessor, Alvaro Uribe, had links to the paramilitaries. Without the benevolence of Abdoulaye Wade, no Grands Domaines du Sénégal! And what would the ever-energetic Jean-Claude Gandur do in Sierra Leone without the corrupt leaders who steal land from rural communities for the sake of his profit?

But corrupt politicians are not the main enemy. It would be absurd and pointless to wait for an awakening of moral conscience among the grain merchants, the vultures of green gold, or the tiger sharks of market speculation. The law of the maximization of profits is a law of iron. But how can we fight and conquer this enemy?

Che Guevara liked to cite this Chinese proverb: "The strongest walls crumble, starting from the cracks." So we must put as many cracks as possible in the current global order that is crushing the world's people, burying them beneath a mantle of concrete. Antonio Gramsci wrote in a letter from prison, "Pessimism of the intellect, optimism of the will." Charles Péguy, whose writings provide the epigraph with which I begin this book, spoke of "hope, this flower of creation" that "dazzles God himself."

Rupture, resistance, the support of the world's peoples in opposing the established order are indispensable, at every level—globally and locally, in theory and in practice, here and everywhere. There must be concrete, voluntary acts, like those in which the farmers' unions in Ross Béthio, Benin, or in the high sierra of the Yucatán in Guatemala, or in the rice fields of Las Pavas, Colombia, are engaged.

In parliaments, in international regulatory authorities, we can decide that there must be change; we can decide to make the right to food a priority, to remove food from the realm of market speculation, to protect subsistence agriculture in the name of national heritage and invest in improving it worldwide. The solutions exist; the plans and projects are already drafted. What is lacking is the will of governments.

And yet, in the West at least, by voting, by taking advantage of our rights to free expression, by mobilizing on a large scale, by going on strike if necessary, we can obtain radical change in alliances and in policies. There is no powerlessness in democracy.

Today, there is a war being fought between the manioc fields and the sugarcane plantations, between family subsistence agriculture and the agro-industrial corporations—a war without mercy. Everywhere—in Central America and at the foot of the volcanoes along the equator, in the Sahel in southern Africa, on the plains of Madhya Pradesh and Orissa in India, in the Ganges

River delta in Bangladesh—farmers, stockbreeders, herders, and fishers are mobilizing, organizing, resisting.

The global reign of the agro-industrial conglomerates creates scarcity, famine for hundreds of millions of human beings, death. Family subsistence agriculture, by contrast, so long as it is supported by governments and so long as farmers can acquire the necessary investments and inputs, is a guarantor of life—for all of us.

The preamble of the declaration presented by La Via Campesina to the UN Human Rights Council during its sixteenth session, in March 2011, warns us solemnly:

> Almost half of the people in the world are peasants. Even in the high-tech world, people eat food produced by peasants. Small-scale agriculture is not just an economic activity; it means life for many people. The security of the population depends on the well-being of peasants and sustainable agriculture. To protect human life it is important to respect, protect and fulfill the rights of the peasants. In reality, the ongoing violations of peasants' rights threaten human life.

Nothing less is required than our total solidarity with the millions of human beings whom hunger is destroying. As the words of a magnificent song made famous by Mercedes Sosa implore:

> *Sólo le pido a Dios*
> *que el dolor no me sea indiferente,*
> *que la reseca muerte no me encuentre*
> *vacío y solo, sin haber hecho lo suficiente.*

> I only ask of God
> That I not be indifferent to suffering,
> That parched death not find me
> Empty and alone, without having done enough.

NOTES

Translator's note on sources: In the original French edition of this book, Jean Ziegler quotes from sources published in many languages, almost always in French translation, sometimes his own. For sources originally published in English, I have quoted from the English original wherever possible. In the cases of a few sources originally written in English to which I could not gain access, I have preferred to transform Ziegler's own French translations of quoted passages into paraphrase rather than translating them in effect back into English via French. For sources in other languages, I have quoted wherever possible from previously published or otherwise official English translations. All translations from French sources not available in English translation are my own. For sources in languages other than French and English, in a very few instances, when I could not locate a citation in its original language (and make my own translation) or in a published English translation, I have had no choice but to translate brief quotations directly into English from Ziegler's French text; this might mean in effect that I have made a translation via French from, for example, a Portuguese or Dutch original. Such instances are generally pointed out in the notes below.

I have used "Ibid." only when the source referred to is entirely unambiguous. Subsequent references to a source after its initial citation are in short form by author's last name and short title, with a cross reference to the first, complete, citation in the notes.

My additions within the author's notes (enclosed within square brackets) and new notes that I have felt it necessary or useful to add for the English-language reader are all marked: "—Trans."

The French edition of this book includes very few online references for the documents cited. For the reader's convenience, I have added these wherever possible; URLs included without further comment are not tagged as my additions. This book is clearly intended not only to provide information but to inspire action. I hope that readers will find the online resources included below useful.

I wish to express my gratitude to editorial director Marc Favreau and to editors Azzurra Cox and Sarah Fan for their advice and encouragement, and to copy editor Sue Warga and proofreader Susan Barnett for their sharp eyes. Above all, I thank executive director Diane Wachtell for the opportunity of helping to bring this important book into English, and express my solidarity with Jean Ziegler and his collaborators in their mission.

—C.C.

PREFACE

xiv   *Geopolítica da fome*   Josué de Castro, *Geopolítica da fome* (Rio de Janeiro: Casa do Estudante do Brasil, 1951). This book was ultimately translated into twenty-six languages; however, it was originally published in the United States with the confusing—and inappropriate—title *The Geography of Hunger* (Boston: Little, Brown, 1952). The translation's title was clearly borrowed from De Castro's earlier pioneering work on the subject, *Geografia da fome: o dilema brasileiro: pão ou aço* (Rio de Janeiro: O Cruzeiro, 1946, with several revised editions through 1971; the subtitle means, "Or, The Brazilian Dilemma: Bread or Steel?"), which provided the foundation for the 1951 book. A new, more appropriately titled English edition was later published as *The Geopolitics of Hunger* (New York: Monthly Review Press, 1977); this later English edition also includes all the revisions and additions de Castro made for the French editions of the book in 1964 and 1973. In order to prevent confusion between these two of de Castro's works and between the later book's two different English-language editions, I have used the Portuguese titles throughout the text except when referring specifically to one of the translations. —Trans.

xiv   *whose article 25 . . . nutrition*   The relevant clause in article 25 reads: "Everyone has the right to a standard of living adequate for the health and well-being of himself and of his family, including food, clothing, housing and medical care. . . ." —Trans.

xv   *Without these young scholars . . . possible*   For further information, visit our website: www.rightfood.org.

xv   *In 2009, there were . . . Geneva*   See Blaise Lempen, *Genève, Laboratoire du XXIe siècle* (Geneva: Éditions Georg, 2010).

xvii   *alienated by the doxa*   The term *doxa*, widely used by French sociologists, anthropologists, and cultural theorists, has no exact equivalent in English.

Adapting the term from Greek philosophy and rhetoric, in which it meant "common belief" or "popular opinion," Pierre Bourdieu, in his *Esquisse d'une théorie de la pratique* (1972; translated as *Outline of a Theory of Practice*, 1977) redefined the term to mean all that is taken for granted and unquestioned in a particular society. Thus the doxa encompasses everything in the individual's experience that makes the natural and the social world appear to be self-evident and defines not only what is sayable but what is even thinkable. The concept of the doxa refines the Marxian idea of alienation: in effect, the doxa limits the possibilities of social, political, and economic change by limiting individuals' capacity to conceive of a changed social order. The paradox of the term lies in the user's implicit assertion of a unique exemption from the doxa's coercive power. Where it occurs elsewhere in this book, the term is left as such, since related English words such as *ideology* and *orthodoxy* do not adequately render its meaning. —Trans.

xvii *"It is not . . . approve of it"* Max Horkheimer, preface to the second edition of *Théorie traditionnelle et théorie pratique*, trans. Claude Maillard and Sybille Muller (Paris: Éditions Gallimard, 1974), 10–11. [For an in-print volume of Horkheimer's essays from the 1930s that largely matches the contents of this selection in French and was published in the same period, immediately after his death, see Max Horkheimer, *Critical Theory: Selected Essays*, trans. Matthew J. O'Connell (New York: Continuum, 1975).]

xvii *"anticipatory consciousness"* See Ernst Bloch, *Das Prinzip Hoffnung* (Frankfurt: Suhrkamp, 1959). [Available in English as *The Principle of Hope*, trans. Neville Plaice, Stephen Plaice, and Paul Knight (Oxford: Basil Blackwell, 1986). Bloch's term in German is *vorgelagertes Bewusstsein*. —Trans.]

1. THE GEOGRAPHY OF HUNGER

3 *the International Covenant . . . Rights* The convention was adopted by the UN General Assembly on December 16, 1966. [However, it did not enter into force until January 3, 1976. —Trans.]

3 *The right to . . . fear* See the website of the United Nations Special Rapporteur on the Right to Food: www.srfood.org/index.php/en/right-to -food. —Trans.

3 *"A meagre diet . . . blood"* Ecclesiastes 34:21–22. [This translation, from the New Jerusalem Bible (www.catholic.org/bible/book.php?id=28) is closest to the standard French Jerusalem Bible version cited by the author. —Trans.]

4 *The fundamental unit . . . itself* The term *calorie* is intrinsically ambiguous and can be confusing. The familiar *dietary, food,* or *nutritionist's calorie* is technically termed the *kilocalorie* (also called the *kilogram calorie* or *large calorie*), since it is equivalent to 1,000 *gram calories* or *small calories*. The small calorie, which equals the amount of heat energy needed to raise the temperature of 1 gram of water by 1 degree Centigrade at sea level, is much too small to be useful in nutritional contexts. —Trans.

6 *For the poor . . . force* Peter Piot, *The First Line of Defense: Why Food and Nutrition Matter in the Fight Against HIV/AIDS* (Rome: World Food Program Public Affairs Service Resources Department, 2004), 4. [This twelve-page brochure is available in five languages at www.wfp.org/content/first-line -defence. —Trans.]

6   *In Switzerland . . . hunger*   National Demographic Institute, Paris, 2009.

7   *The mathematical model . . . complex*   Regarding the FAO's statistical modeling, I have benefited from the invaluable assistance of Pierre Pauli, a statistician in Geneva's Office Cantonal de la Statistique.

7   *tens of millions of metric tons of wheat*   Throughout this book, I have retained specific measurements in the SI, or metric, system naturally used by Jean Ziegler, adding U.S. equivalents in parentheses. With one exception: since measurements given in metric tons throughout are always very large and somewhat approximate, and since the metric ton and the U.S./imperial long and short ton are all very close in mass, I have simply used *ton* to mean metric ton throughout, and refrained from providing a non-SI equivalent. —Trans.

8   *Bernard Maire . . . model*   See Francis Delpeuch and Bernard Maire, *Alimentation, environnement et santé: Pour un droit à l'alimentation* (Paris: Éditions Ellipses, 2010).

9   *Latest available statistics . . . 2010)   Global Hunger Declining, But Still Unacceptably High: International Hunger Targets Difficult to Reach* (Rome: FAO, Economic and Social Development Department, 2010), www.fao.org/docrep/012 /al390e/al390e00.pdf. See also *The State of Food Insecurity in the World: Addressing Food Security in Protracted Crises* (Rome: FAO/WFP, 2010), www .fao.org/docrep/013/i1683e/i1683e00.htm, and the FAO Hunger Portal: www.fao.org/hunger.

11   *The International Fund . . . earth*   See IFAD, *Rural Poverty Report 2009* (New York: Oxford University Press, 2010).

11   *In view of . . . simple*   See Jean Feyder, *Mordshunger: Wer profitiert vom Elend der armen Länder?* (Munich: Westend, 2010). [This book has been translated into French as *Faim tue* (Paris: L'Harmattan, 2011) but is not yet available in English. —Trans.]

12   *Anyone who has . . . admiration*   The organization is also known more simply as the Movimento dos Trabalhadores Sem Terra (Landless Workers' Movement). —Trans.

12   *food sovereignty*   "*Food sovereignty* is a term coined by members of Via Campesina in 1996 to refer to a policy framework advocated by a number of farmers, peasants, pastoralists, fisherfolk, indigenous peoples, women, rural youth and environmental organizations, namely the claimed 'right' of peoples to define their own food, agriculture, livestock and fisheries systems, in contrast to having food largely subject to international market forces" (Wikipedia, "Food Sovereignty," accessed March 18, 2012). —Trans.

14   *In rural areas . . . 2005*   See UN Commission on Human Rights, *Report of the Special Rapporteur on the Right to Food: Addendum, Mission to Guatemala*, E/CN .4/2006/44/Add.1 (United Nations Economic and Social Council, 2006), available at www.unhcr.org/refworld/docid/4411820a0.html.

16   *As for the violent crimes . . . increased*   See *The Right to Food in Guatemala: Final Report of the International Fact-Finding Mission* (Guatemala City: Food Information and Action Network [FIAN], 2010), www.fian.org/fileadmin /media/publications/2010_03_Guatemala_FactFindingMission.pdf.

17   *Manila's Smokey Mountain landfill*   Smokey Mountain is a forty-year-old municipal waste dump near Manila, where some thirty thousand people live, surviving by scavenging amid the garbage. —Trans.

17 *The Nicaraguan government . . . $80)* Yolanda Areas Blas, remarks during a side event titled "The Need of Increased Protection of Human Rights of Peasants" at the sixteenth session of the Human Rights Council, Geneva, March 9, 2011. See Jean Ziegler, *Update on the Preliminary Study of the Human Rights Council Advisory Committee on the Advancement of the Rights of Peasants and Other People Working in Rural Areas*, A/HRC/16/63 (Geneva: Human Rights Council Advisory Committee, 2011), www2.ohchr.org/english/bodies /hrcouncil/docs/16session/A-HRC-16-63.pdf.

17 *The geographical distribution . . . unequal* All of the graphs and tables in this chapter are adapted from *The State of Food Insecurity in the World: Addressing Food Security in Protracted Crises* (Rome: FAO/WFP, 2010), www.fao.org/docrep /013/i1683e/i1683e00.htm.

22 *another 25 million . . . "in transition"* The situation is especially serious in many orphanages where, according to some American NGOs, employees allow children to starve to death. One example: in the orphanage in Torez, in Ukraine, 12 percent of the children, most of them disabled, die of hunger every year. See Daniel Foggo and Martin Foley, "Ukrainian Orphanages 'Are Starving Disabled Children,'" *Sunday Times* (London), February 6, 2011. [This article appears to be inaccessible on the *Times*'s website but can be read here: www.hiskidstoo.org/news/article-from-sunday-times -london. —Trans.]

23 *Ever since . . . French protectorate* Officially a "protectorate," Tunisia was effectively a French colony from the French invasion in 1881 until independence in 1956. —Trans.

24 *appalling custom . . . do not eat* "Affamati, ma a casa loro," *Nigrizia*, July 1, 2009, www.nigrizia.com/sito/notizie_pagina.aspx?Id=7872&IdModule =1. [This article is not available in English. —Trans.]

2. INVISIBLE HUNGER

25 *Several million children . . . year* Hans Konrad Biesalski, "Micronutriments, Wound Healing and Prevention of Pressure Ulcers," *Nutrition* 28 (September 2010): 858.

26 *Micronutrient Initiative . . .* Report See UNICEF/The Micronutrient Initiative, *Vitamin and Mineral Deficiency: A Global Damage Assessment Report* (2004). The complete range of MI's reports is available at www.micronutrient .org/English/View.asp?x=614&id=473. —Trans.

27 *About 30 percent . . . life* See Hartwig de Haen, "Das Menschenrecht auf Nahrung," conference presentation, Einbeck-Northeim, January 28, 2011. [Although this speech is not available in English, other papers and interviews by de Haen, the former assistant director general of the FAO's Economic and Social Department, are accessible online, including a 2002 interview on food insecurity available at www.fao.org/english/newsroom /news/2002/9703-en.html. —Trans.]

27 *According to a study . . . annually* "Hidden Hunger: How Much Can Farming Really Improve People's Health?" *The Economist*, March 24, 2011, www .economist.com/node/18438289.

27 *Zinc deficiency also . . . children* Nicholas D. Kristof, "Bless the Orange Sweet

Potato," op-ed column, *New York Times*, November 24, 2010, www.nytimes
.com/2010/11/25/opinion/25kristof.html.

28   *"To end childhood malnutrition . . . will"*   *En finir avec la malnutrition, une question de priorité*, proceedings of the conference of the same title (Paris: Action Contre la Faim, 2008). [This document appears to be no longer available online; for information on the organization's more recent work, see www .actioncontrelafaim.org/en. The website of the U.S. branch is www.action againsthunger.org. —Trans.]

3. PROTRACTED CRISES

29   *A country thus afflicted . . . inhabitants*   See Paul Collier, *The Bottom Billion: Why the Poorest Countries Are Failing and What Can Be Done about It* (London: Oxford University Press, 2008).

30   *All the countries . . . disasters)*   For a detailed list of the criteria that govern inclusion on the list of LDCs, see "The Criteria for the Identification of the LDCs," www.un.org/special-rep/ohrlls/ldc/ldc%20criteria.htm; the current list of LDCs is available at www.unohrlls.org/en/ldc/25. —Trans.

32   *The royalties paid . . . low*   Greenpeace Switzerland, dossier on Areva/Niger released at a press conference, Geneva, May 6, 2010.

33   *migrating desert locusts*   The desert locust, *Schistocerca gregaria Forsk.*, is called the *criquet pèlerin* in French, a term occasionally rendered literally in English as "pilgrim locust." —Trans.

35   *The locusts invaded . . . Egypt*   Exodus 10:14–15, New Jerusalem Bible.

38   *In February 2005 . . . control*   See *Report of the United Nations Fact Finding Mission on the Gaza Conflict* (New York: United Nations Human Rights Council, 2009), www2.ohchr.org/english/bodies/hrcouncil/specialsession/9/Fact FindingMission.htm. Established by the UNHRC during the 2009 Gaza War, the mission was headed by South African jurist Richard Goldstone. Hereinafter I will refer to this document as "the Goldstone report" [as it is commonly known. —Trans.]. In 2011, Goldstone, under intense pressure from Jewish groups (Goldstone is himself Jewish), attempted to modify some of the report's original conclusions. A majority of the commissioners prevented this from happening.

38   *As an occupying power . . . population*   See the reports of Richard Falk, the UN Special Rapporteur on the Occupied Palestinian Territories, particularly his reports dated June 2010 (A/HRC/13/53/Rev. 1; available at unispal.un.org /UNISPAL.NSF/0/33F2A0A73AB185DB8525773E00525D05), August 2010 (A/HRC/65/331; available at unispal.un.org/UNISPAL.NSF/0/6 9BEC99AF727EAC2852577C3004AAD8A), and January 2011 (A/HRC /16/72; available at unispal.un.org/UNISPAL.NSF/0/A72012A31C1116E C8525782C00547DD4).

38   *Step by step . . . UNRWA*   AbuZayd, appointed by UN General Secretary Kofi Annan, served as commissioner-general from June 28, 2005, to January 20, 2010; she was succeeded by her deputy, Filippo Grandi. —Trans.

39   *The entire population . . . responsibility*   See International Committee of the Red Cross, "Gaza Closure: Not Another Year!" June 14, 2010, www.icrc .org/eng/resources/documents/update/palestine-update-140610.htm. See

also Christophe Oberlin, *Chroniques de Gaza, 2001–2011* (Paris: Éditions Demi-Lune, 2011).

39 *On December 27 . . . limbs* See the Goldstone report, chap. 6, sec. C, "Data on casualties during the Israeli military operations in Gaza from 28 December 2008 to 17 January 2009." Among the Israeli forces, there were ten casualties, many due to friendly fire.

39 *The attacking Israeli forces . . . destroyed* See the Goldstone report, chap. 13, "Attacks on the Foundations of Civilian Life in Gaza: Destruction of Industrial Infrastructure, Food Production, Water Installations, Sewage Treatment Plants and Housing" [especially sec. A, "The Destruction of el-Bader Flour Mill" —Trans.].

40 *two successive attacks . . . water* See the Goldstone report, chap. 13, sec. C, "The Destruction of Water and Sewage Installation," part 1, "The Gaza Wastewater Treatment Plant, Road No. 10, al-Sheikh Ejlin, Gaza City." The missile strike caused massive flooding of farmland with raw sewage. The report concludes: "The fact that the lagoon wall was struck precisely . . . where it would cause outflow of the raw sewage suggests that the strike was deliberate and premeditated." —Trans.

40 *Richard Goldstone, president . . . targets* See the Goldstone report, chap. 11, "Deliberate Attacks Against the Civilian Population." See also Mads Gilbert and Erik Fosse, *Eyes in Gaza* (London: Quartet Books, 2010).

40 *Stéphane Hessel and . . . weapon* Stéphane Hessel and Michel Warschawski, speeches at the conference titled Crimes de Guerre, Blocus de Gaza, held at the University of Geneva, March 13, 2011. [Video footage of Hessel's and Warschawski's remarks at the conference (in French) is available from many online sources, including www.genevelibertaire.ch/~taz/archives/1245. —Trans.]

41 *the manhunts undertaken . . . diaspora* See Juliette Morillot and Dorian Malovic, *Évadés de Corée du nord* (Paris: Belfond, in the Témoignages series, 2004). This book comprises eyewitness accounts collected in Manchuria and South Korea from survivors of the North Korean famine. [Not available in English. —Trans.]

41 *From 1996 to 2005 . . . people* See Jacques Follorou, "Le système de répression politique en Corée du Nord s'est renforcé et aggravé," *Le Monde*, May 12, 2011; and Philippe Pons, "Corée du Nord: Pénurie alimentaire et jeu politique," *Le Monde*, May 14, 2011. These articles are available to subscribers or for purchase at, respectively, www.lemonde.fr/cgi-bin/ACHATS/acheter.cgi?offre=ARCHIVES&type_item=ART_ARCH_30J&objet_id=1156681 and www.lemonde.fr/cgi-bin/ACHATS/acheter.cgi?offre=ARCHIVES&type_item=ART_ARCH_30J&objet_id=1157024.

41 *The Kim dynasty . . . hunger* Kim Il-sung, the founder of the dynasty, ruled North Korea, officially the Democratic People's Republic of Korea, from its establishment in 1948 until his death in 1994. He was succeeded by his son Kim Jong-il, who was in turn succeeded by *his* son Kim Jong-eun in 2011. —Trans.

42 *Amnesty International . . . freed* "Images Reveal Scale of North Korean Political Prison Camps," Amnesty International news release, May 3, 2011, www

.amnesty.org/en/news-and-updates/images-reveal-scale-north-korean-political-prison-camps-2011-05-03. See also Amnesty International, *Annual Report 2011: The State of the World's Human Rights—North Korea*, www.amnesty.org/en/region/north-korea/report-2011.

42 *Many returned famine refugees... for life*   "End Horror of North Korean Political Prison Camps," Amnesty International call for action, May 4, 2011, www.amnesty.org/en/appeals-for-action/north-korean-political-prison-camps. See also Morillot and Malovic, *Évadés de Corée* (see note to p. 41).

42 *Amnesty International... months*   Jack Rendler, "North Korean Prison Camps Grow Larger," *Human Rights Now*, Amnesty International USA blog, May 11, 2011, blog.amnestyusa.org/iar/north-korea-the-last-worst-place-on-earth.

42 *up to 40 percent... feces*   See Amnesty International, "Images Reveal Scale" (see note to p. 42): "According to the testimony of a former detainee at the revolutionary zone in the political prison camp at Yodok, an estimated 40 per cent of inmates died from malnutrition between 1999 and 2001. . . . Food in the camps is scarce. Amnesty International has been told of several accounts of people eating rats or picking corn kernels out of animal waste purely to survive, despite the risks—anyone caught risks solitary confinement or other torture." —Trans.

4. THE CHILDREN OF CRATEÚS

43 *Brazil's northeastern states . . . miles)*   The *sertão* comprises parts of seven Brazilian states: Alagoas, Bahia, Pernambuco, Paraíba, Rio Grande do Norte, Ceará, and Piauí.

44 *If he is lucky . . . midday*   The idiomatic English translation of *bóias-frias* is "cold lunchers," though the term's literal meaning is "cold buoys." *Bóia* ("buoy")—can refer to other floating things, such as pieces of bread or meat in a bowl of soup, and thus bread/meat/food in general. —Trans.

44 *And there we . . . bishopric*   Like all the great dioceses of Brazil, the diocese of Crateús has a sumptuous episcopal palace. But upon his nomination in 1964, Dom Fragoso refused to live there. Born in a market town in the interior in the state of Paraíba, Dom Fragoso died in 2006 at the age of eighty-two.

45 *listening to Radio Tirana . . . revolution*   In the Enver Hoxha period [Hoxha ruled Albania as a Stalinist dictator from the end of World War II until his death in 1985. —Trans.], Radio Tirana, the Albanian national radio network, was broadcast on shortwave literally around the world in many languages, including Portuguese.

5. GOD IS NOT A FARMER

47 *The Berne Declaration... aid*   Berne Declaration, bulletin, February 1, 2009.

47 *the first of... hunger*   For details on the first of the Millennium Development Goals, see www.un.org/millenniumgoals/poverty.shtml.

48 *In a study . . . 1:30*   *Average Yield of Rainfed Crops and Irrigated Crops* (Geneva: World Meteorological Organization, 2006). [Although this document is not available online, the organization's more recent publications are available at www.wmo.int/pages/index_en.html. —Trans.]

50 *The title of... Farmer*   Mamadou Cissokho, *Dieu n'est pas un paysan* (Paris:

Présence Africaine, 2009). [This book has not yet appeared in English translation. —Trans.]

6. "NO ONE GOES HUNGRY IN SWITZERLAND"

51 *Jean-Charles Angrand . . . attained"* Jean-Charles Angrand, personal letter to the author, December 26, 2010. [The island of Réunion is a French "overseas department" (a region of France with the same administrative status as the country's mainland *départments*) situated in the Indian Ocean east of Madagascar, with a population of about 800,000. —Trans.]

52 *"Mothers and babies . . . matters* François Soudan, "Les Femmes et les Enfants en Dernier," *Jeune Afrique*, February 8, 2010, www.jeuneafrique.com /Article/ARTJAJA2586p003-004.xml0.

53 *The G8 and the G20 . . . G8+5* The Group of Eight (G8), founded in 1975 as the G6, is an annual forum where the governments of eight of the world's largest economies are represented: France, Germany, Italy, Japan, the United Kingdom, the United States, Canada, and Russia. The European Union sends a representative to the G8 but cannot host or chair summits. When Brazil, China, India, Mexico, and South Africa are represented as guests, the group is known as the G8+5. The Group of Twenty Finance Ministers and Central Bank Governors (G20), which includes representatives from the EU and a wider array of countries that together account for about 80 percent of world GNP and world trade and two-thirds of the world's population, held semi-annual summit meetings from 2008 through 2011. The G20 will meet annually starting in 2012, and has announced that it will essentially supersede the G8. —Trans.

53 *In his memoir . . . career* Tony Blair, *A Journey: My Political Life* (London: Hutchinson, 2010; New York: Vintage, 2011). I refer here to the German edition of Blair's memoir, *Mein Weg*, trans. Helmut Dierlamm et al. (Munich: Bertelsmann, 2010), 623.

54 *As for the promises . . . 2009* "At their meeting in L'Aquila in July, the Group of Eight (G8) large rich economies promised to increase spending on agricultural development by $20 billion over the next three years. Not much of this was new money (probably $3 billion–5 billion) and it is not clear how much, if any, has been delivered." "If Words Were Food, Nobody Would Go Hungry," *The Economist*, November 19, 2009, www.economist.com/node/14926114.

7. THE TRAGEDY OF NOMA

55 *Noma's technical name* *Noma* comes from an idiomatic use of the Ancient Greek verb *nemein*, "to spread" (of an ulcer). See *Merriam-Webster's Collegiate Dictionary*, 11th ed. (Springfield, MA: 2003). —Trans.

56 *One important expert . . . Federation* Winds of Hope (www.windsofhope.org /en) was founded by Bertrand Piccard and Brian Jones in 1999, after they completed the first-ever nonstop around-the-world flight in a balloon, to combat forgotten childhood diseases. The No-Noma International Federation (www.nonoma.org/index.php) was founded as an initiative of Winds of Hope in 2003; Piccard is the president of No-Noma.

57 *at least 80 percent . . . death* According to Facing Africa, a U.K. charity that

helps noma sufferers in Ethiopia, the mortality rate for noma ranges even higher: up to 90 percent. See "What Is Noma?" at www.facingafrica.org /FA08/content/site/en/pages/whatisnoma/default.asp. —Trans.

57 *In most traditional societies . . . committed* Ben Fogle, in the documentary *Make Me a New Face: Hope for Africa's Hidden Children*, broadcast over BBC channels several times June 9–July 5, 2010; for information, see www.bbc.co.uk /programmes/b00sqlrg.

57 *Organisation of the Islamic Conference* The Organisation of the Islamic Conference officially changed its name to the Organisation of Islamic Cooperation on June 28, 2011. —Trans.

58 *Sentinelles* Sentinelles, founded in 1980 by Edmond Kaiser to "come to the aid of wounded innocence," is based in Lausanne: www.sentinelles.org.

58 *Two Dutch plastic . . . possible* Klaas Marck, "A History of Noma, the 'Face of Poverty," *Plastic and Reconstructive Surgery* 111, no. 5 (April 15, 2003): 1702–7; doi:10.1097/01.PRS.0000055445.84307.3C. This article is available for purchase at journals.lww.com/plasreconsurg/Abstract/2003/04150/A _History_of_Noma,_the_Face_of_Poverty_.18.aspx. [Bos and Marck's book, referred to in the next note, also includes a short history of the disease. —Trans.]

58 *They have learned . . . work* See Kurt Bos and Klaas Marck, *The Surgical Treatment of Noma* (Amsterdam: Dutch Noma Foundation and Facing Africa, 2006), www.noma.nl/downloads/nomasurgicaltreatmentpdf.pdf.

59 *Every year, some . . . born* See Cyril O. Enwonwu, "Noma—The Ulcer of Extreme Poverty," *New England Journal of Medicine* 354 (January 19, 2006): 221–24, www.nejm.org/doi/full/10.1056/NEJMp058193.

60 *Medical teams from . . . Laos* M. Leila Srour et al., "Noma in Laos: Stigma of Severe Poverty in Rural Asia," *American Journal of Tropical Medicine and Hygiene* 78, no. 4 (April 2008): 539–42, www.ajtmh.org/content/78/4/539.

61 *noma is present . . . Southeast Asia* There are at present no reliable statistics on the incidence of noma in Asia.

61 *the African regional bureau . . . the World Bank"* See Alexander Fieger et al., "An Estimation of the Incidence of Noma in North-west Nigeria," *Tropical Medicine and International Health* 8, no. 5 (May 2003): 375–483, available at onlinelibrary.wiley.com/doi/10.1046/j.1365-3156.2003.01036.x/full. Based on particular statistical inferences, Fieger and his coauthors estimate the global incidence of noma at only 30,000–40, 000. They continue:

Although this number is lower than previous estimations of the WHO (140 000), the concomitant global mortality is well comparable with that of diseases like trypanosomiasis, leishmaniasis, acute upper respiratory infections, obstructed labour, multiple sclerosis and appendicitis. All these disorders are present in the yearly WHO reports in contrast with noma. This reflects the lack of a good monitoring system for noma, and the lack of interest for this affection [i.e., bodily condition; disease, malady —Trans.] from the side of public health policy makers both in less privileged countries where noma is prevalent and also of mondial institutions like the WHO and the World Bank.

Though the combat against extreme poverty is one of the current targets of the World Bank, it is not realized that an excellent biological parameter for the presence of extreme poverty in a population is the structural presence of noma. Noma completely disappears from a society if economical progress enables the poorest in their society to feed their children sufficiently. —Trans.

61 *Many publications issued . . . noma* The WHO has partnered with many other organizations, including the World Bank, on the Global Burden of Disease (GBD) project since 1990, when the first GBD study quantified the health effects of more than a hundred diseases and injuries for eight regions of the world. WHO is currently collaborating with the Institute for Health Metrics and Evaluation and other academic partners in a new round of the GBD 2010 study for the years 1990 and 2005, the Global Burden of Diseases, Injuries, and Risk Factors Study 2010 Health Measurement Survey. All the GBD publications and their periodic updates are available at www.who.int /healthinfo/global_burden_disease/en. —Trans.

62 *Swiss Federal Office of Public Health* Information in English on this Swiss agency is available at www.bag.admin.ch/org/index.html?lang=en. —Trans.

62 *Realizing this action plan . . . have* See Olivier Grivat, "Notre but: Mettre sur pied une journée mondiale contre le noma," an interview with Bertrand Piccard, *Tribune Médicale* 39 (September 29, 2006), available at www .nonoma.org/index.php?option=com_alphacontent&itemid=51&lang =French (under the heading "Dernières nouvelles").

8. FAMINE AND FATALISM: MALTHUS AND NATURAL SELECTION

66 *his famous work* The complete title of Malthus's essay in its anonymously published first edition (1798) was *An Essay on the Principle of Population, as it affects the future improvement of society with remarks on the speculations of Mr. Godwin, M. Condorcet, and other writers.* In the second, much enlarged edition (1803), which acknowledged Malthus's authorship, the subtitle was changed: *An Essay on the Principle of Population; or, a view of its past and present effects on human happiness; with an enquiry into our prospects respecting the future removal or mitigation of the evils which it occasions.* The new subtitle was retained in the third through sixth editions. —Trans.

66 *Malthus revised the work . . . revising* Most of Malthus's substantive revisions to the *Essay* were made for the second edition of 1803; in the subsequent editions (1806, 1807, 1817, and 1826), most of Malthus's changes were refinements of diction. —Trans.

67 *Through the animal . . . vice* Thomas Malthus, *An Essay on the Principle of Population* (1798), chap. 1. The *Essay* is available in several formats from Project Gutenberg, www.gutenberg.org/ebooks/4239.

67 *Under this law . . . rent* Thomas Robert Malthus, *Principles of Political Economy*, 2nd ed., published posthumously (London: W. Pickering, 1836), chap. 3, "Of the Rent of Land," sec. 9, "General Remarks on the Surplus Produce of the Land"; the entire book is available at oll.libertyfund.org

/index.php?option=com_staticxt&staticfile=show.php%3Ftitle=2188&Ite
mid=99999999.

68 *the Poor Laws* The English Poor Laws had their roots in late medieval law
and were first codified at the end of the sixteenth century. The system,
which was drastically overhauled by the New Poor Law of 1834, was the
subject of considerable controversy in Malthus's lifetime. —Trans.

69 *"The tribes of hunters . . . other"* Thomas Robert Malthus, *An Essay on the
Principle of Population*, 6th ed. (London: John Murray, 1826), book 1, chap. 4,
"Of the Checks to Population among the American Indians," available at
www.econlib.org/library/Malthus/malPlong2.html#Chapter%20IV.

9. JOSUÉ DE CASTRO, PHASE ONE

70 *In his book . . . sense* Ludwig Feuerbach, *Das Wesen des Christenthums* (1841;
2nd ed., 1848). One of the most celebrated translations of this book is by the
British novelist George Eliot, writing under her real name, and it remains
standard: *The Essence of Christianity*, 1st ed., trans. from the 2nd German ed.
by Marian Evans (London: John Chapman, 1854). The passage to which
Jean Ziegler refers appears on pages 1–2:

> Consciousness in the strictest sense is present only in a being to whom
> his species, his essential nature, is an object of thought. . . . Where
> there is this higher consciousness there is a capability of science.
> Science is the cognizance of species. In practical life we have to do
> with individuals; in science, with species. But only a being to whom
> his own species, his own nature, is an object of thought, can make the
> essential nature of other things or beings an object of thought.
> —Trans.

71 *two agreements . . . Rights* The complete texts of these two covenants are
available, respectively, at www2.ohchr.org/english/law/cescr.htm and
www2.ohchr.org/english/law/ccpr.htm.

72 *Garota da Ipanema* The restaurant's name means "The Girl from Ipanema";
it memorializes the 1962 hit bossa nova song of the same name with music by Antônio Carlos Jobim and Portuguese lyrics by Vinícius de Moraes.
—Trans.

73 Geografia da fome Josué de Castro, *Geografia da fome* (Rio de Janeiro: O
Cruzeiro, 1946). Although it was translated into French in 1949 and reissued in 1964, this pioneering work appears never to have been translated
into English. [See note to p. xiv. —Trans.]

74 *"The table of . . . fecund"* The French version of this maxim appears as an
epigraph in the French edition of de Castro's *Geopolítica da fome* (see note to
p. 75). —Trans.

74 *"If some of . . . fuel"* This quotation is translated from Jean Ziegler's
own French version of the passage (see the following note). *Mestiço* is the
Portuguese equivalent of the more familiar Spanish term *mestizo*. Brazilians
use a highly technical vocabulary to denote people of various mixed ethnic
heritage that expresses different attitudes from those familiar to Americans
on the subject of race. For a good brief overview, see the section on "Brazil"
in the Wikipedia entry on the term "Multiracial," en.wikipedia.org/wiki

/Multiracial#Brazil; or for more detail, the article on "Race in Brazil," en.wikipedia.org/wiki/Race_in_Brazil. —Trans.

74 Documentario do Nordeste . . . *race*   The two books discussed here, which have not been translated into French and appear not to have been translated into English either, are *Alimentação e Raça* (Food and Race; Rio de Janeiro: Civilização Brasileira, 1935) and *Documentário do Nordeste* (Documentary of the Northeast; Rio de Janeiro: José Olympio, 1937). The latter should not be confused with a book that *is* available in English, de Castro's *Death in the Northeast: Poverty and Revolution in the Northeast of Brazil* (New York: Vintage, 1969), which is a translation of a much later work, *Sete palmos de terra e um caixão* (literally, Five Handspans of Earth and a Coffin; São Paulo: Brasiliense, 1965). —Trans.

75 *In 1945 . . . Novo*   Vargas returned to power in 1951 [in a democratic election —Trans.], but, discredited and on the verge of being driven from power, he committed suicide in 1954 by shooting himself in the heart in the presidential Catete Palace in Rio de Janeiro [today a museum —Trans.].

75 *his scientific work . . . fifty titles*   About half of de Castro's works have been translated into other major languages.

75 *"The central argument . . . disabled'"*   Alain Bué, "La tragique nécessité de manger," *Politis*, October–November, 2008. Bué was de Castro's assistant at the Centre Universitaire Expérimental in Vincennes (later the Université de Vincennes), founded in 1968. Today a professor at the Université de Paris VIII, he is de Castro's intellectual heir and guardian of his legacy in France.

75 *Geografia da fome . . . Geopolítica da fome*   For a clarification of the problematic titles of the English-language translations of de Castro's works, see preface, note to p. xiv. —Trans.

75 *In his author's preface . . . fome*   Josué de Castro, *Géopolitique de la faim*, trans. Léon Bourdon (Paris: Économie et Humanisme and Les Éditions Ouvrières, 1952; revised and expanded edition, 1965), 20. [This author's preface is not included in the 1977 English edition of the book and has not been translated into English. —Trans.] The association of Christian economists known as Économie et Humanisme was founded in 1941 in Marseilles by the Dominican friar Louis-Joseph Lebret. This French translation was thus published on the initiative of a Christian movement that strove in this period especially to reconcile political economy with the Church's social work.

76 *It was in homage . . . book*   The original French edition of this book is titled *Destruction massive: Géopolitique de la faim* (Mass Destruction: The Geopolitics of Hunger). —Trans.

76 *But, although degraded. . . hunger"*   De Castro, *Géopolitique de la faim*, 20–21 (see note to p. 75).

77 *In many of . . . Hunger*   Josué de Castro, *The Black Book of Hunger*, trans. Charles Lam Markmann (New York: Funk and Wagnalls, 1967; Boston: Beacon Press, 1969), a translation of *O livro negro da fome* (São Paulo: Brasiliense, 1957).

77 *capitanias*   During the colonial conquest of Brazil, the king of Portugal gave grants of land to the *fidalgos* (noblemen) comprising sections of the coast, with the provision that they in return conquer the interior. The *figalgo* thus became a *capitão* (captain) and the land that he succeeded in taking

from the indigenous peoples became known as a *capitania* (captaincy) or *doação* (grant). Most of the plantations today are former *capitanias*.

78 *In 1954 . . . exhausted*   Miguel Arrães de Alencar, in conversation with the author.

79 *Tibor Mende worked . . . Shadow*   Although Mende was Hungarian, almost all of his works appear to have been published first in English, but only *China and Her Shadow* (Whitefish, MT: Literary Licensing, 2011; orig. pub. London: Thames and Hudson, 1961) appears to be currently in print. Jean Ziegler also notes *L'Inde devant l'orage* (Paris: Seuil, 1950), ostensibly a translation from an English original, the title of which I cannot determine from the first editions in the New York Public Library or the Bibliothèque Nationale de France; the French title means "India Facing the Storm." Jean Ziegler also notes a work first published in French and not available in English: *Fourmis et poissons: Carnets de route* (Paris: Seuil, 1979). —Trans.

79 *René Dumont's key early books*   René Dumont and Marcel Mazoyer, *Socialisms and Development* (London: Deutsch, 1973), a translation of *Développement et socialisme* (Paris: Seuil, 1969). Jean Ziegler also cites two works not available in English: *Le développement agricole africain* (Paris: Presses Universitaires de France, 1965) and *Paysanneries aux abois* (Paris: Seuil, 1972). —Trans.

79 *the Emmaus movement*   Today's successor to Abbé Pierre's original Emmaus movement is Emmaus International; see emmaus-international.org/index .php?lang=English.

79 *IRFED . . . founded in 1958*   The institute is now known as the Centre International Développement et Civilizations–Lebret-IRFED, based in Paris and Geneva (www.lebret-irfed.org); a related organization is IRFED Europe (Institut International de Recherche et de Formation Education Culture et Développement). —Trans.

80 *Lebret mobilized . . . de Castro*   See especially Louis-Joseph Lebret, *Dimension de la charité* (Paris: Éditions Ouvrières, 1958) and *Dynamique concrète du développement* (Paris: Éditions Ouvrières, 1967). [These books have not been translated into English. —Trans.]

80 *The murder rate . . . world*   Gilliat H. Falbo, Roberto Buzzetti, and Adriano Cattaneo, "Les enfants et les adolescents victimes d'homicide: Une étude cas-témoins à Recife," *Bulletin de l'Organisation Mondiale de la Santé, Recueil d'articles* 5 (Geneva, 2001).

80 *Demetrius Demetrio . . . families*   For information about Pequenos Prophetas, visit www.pequenosprofetas.org.br, which links to the organization's pages on social media sites including Blogspot (in Portuguese only) and Facebook.

81 *In his novel . . . continues*   Josué de Castro, *Of Men and Crabs* (New York: Vanguard, 1970), a translation of *Homens e caranguejos* (Porto: Brasília, 1967).

10. HITLER'S "HUNGER PLAN"

82 *half the population . . . 1942–43*   See Timothy Snyder, *Bloodlands: Europe Between Hitler and Stalin* (New York: Basic Books, 2010).

82 *As soon as Germany . . . satisfied"*   Josué de Castro, *The Geography of Hunger* (Boston: Little, Brown, 1952), 249, quoting Boris Shub and Zorah Warhaftig, *Starvation over Europe (Made in Germany), A Documented Record* (New York: Institute of Jewish Affairs of the American Jewish Congress and World

Jewish Congress, 1943), no page reference. [Facts and figures throughout this chapter are all drawn from de Castro's book; de Castro credits Shub and Warhaftig with much of his data. All subsequent quotations from de Castro's book are from this, the first English edition. —Trans.]

83 *Such was Europe . . . camp* De Castro, *Geography of Hunger*, 252–53.

83 *The "well fed" . . . machine* De Castro notes that "The Germans themselves were really the only well-fed group. . . . The collaborating peoples, who were engaged in tasks of vital or military importance for German security, received a diet that permitted them to maintain a certain degree of labor efficiency." *Geography of Hunger*, 249–50. —Trans.

84 *the Reichsnährstand . . . control* See Clifford R. Lovin, "Agricultural Reorganization in the Third Reich: The Reich Food Corporation (Reichsnahrstand), 1933–1936," *Agricultural History* 43, no. 4 (October 1969): 447–61. —Trans.

84 *Ley declared . . . race"* See de Castro, *Geography of Hunger*, 250–51; de Castro credits Shub and Warhaftig, *Starvation over Europe*, for this information. [Ley's comments were published in *Der Angriff* on January 31, 1940. A fuller citation reads as follows: "A lower race needs less room, less clothing, less food, and less culture than a higher race. The Germans cannot live in the same fashion as the Poles and the Jews. . . . More bread, more clothes, more living space, more culture, more beauty—these our race must have, or it will perish." See "How the Germans Are Starving Poland," Polish Ministry of Information, London, *Polish Fort-Nightly Review* no. 58 (December 15, 1942), available at www.holocaustresearchproject.org/nazioccupation/poland starved.html. —Trans.]

85 *General Government* The full name of the Nazi regime was initially the *Generalgouvernement für die besetzten polnischen Gebiete*, or General Government for the Occupied Polish Territories. In July 1940 the name was officially shortened to Generalgouvernement. A more correct English translation of this term would be "General Governorate"; "General Government," however, has considerable currency in historical writing about this period. The region was unofficially referred to by Germans as *Restpolen*—"the rest of Poland." —Trans.

85 *Curzio Malaparte . . . Kaputt* Curzio Malaparte was a disaffected Italian Fascist sent by an Italian newspaper to cover the war on the Eastern front; *Kaputt*, which grimly depicts the depravity of the Nazi leaders and the horrors of the war, has recently been reissued, translated from the Italian by Cesare Foligno, with an afterword by Dan Hofstadter (New York: New York Review Books, 2005). —Trans.

86 *One day the Germans . . . mouths"* De Castro, *Geography of Hunger*, 251, quoting Else Margrete Roed, "The Food Situation in Norway," *Journal of the American Dietetic Association* 19, no. 12 (December 1943): 817–19.

86 *from 10 to 15 grams . . . 2.5 grams* These figures specifically are for the Hunger Winter of 1944–45 in the Netherlands; see de Castro, *Geography of Hunger*, 254. —Trans.

87 *"the Polish people . . . trees'"* De Castro, *Geography of Hunger*, 253, quoting Maria Babicka, "The Current Food Situation Inside Poland," *Journal of the American Dietetic Association* 19, no. 4 (April 1943).

87 *The Nazi strategy* . . . Hungerplan See Sönke Neitzel and Harald Welzer, "Pardon wird nicht gegeben: Der Krieg gegen die Sowjetunion und die Verbrechen an Kriegsgefangenen," *Blätter für Deutsche und Internationale Politik,* June 2011, 112–23, available for purchase in German only at www .blaetter.de/archiv/jahrgaenge/2011/juni/pardon-wird-nicht-gegeben.

87 *Reichssichertshauptamt* The name of this department in the Nazi security apparatus is also, and more commonly, given as Reichssicherungshauptamt. —Trans.

87 *Thus, everywhere in* . . . *hunger* See Adam Hochschild, "Tug of War: Timothy Snyder Looks East," review of Timothy Snyder, *Bloodlands: Europe Between Hitler and Stalin, Harper's Magazine,* February 2011, 79–82, available to subscribers only at harpers.org/archive/2011/02/page/0081.

88 *Historian Timothy Snyder.* . . . *shot* See Snyder, *Bloodlands* (see note to p. 82).

89 *Journalist Max Nord* . . . *heed"* Max Nord, introduction to *Amsterdam tijdens de Hongerwinter* (Amsterdam, 1947). [This rare and treasured book today commands prices from $400 to $5,000. More background about the volume is available at sites.google.com/site/bintphotobooks/amsterdamtijdensde hongerwinter; a somewhat crude "slide show" of some of the book's images is available at www.youtube.com/watch?v=h1vdcO9BtTU. The first and third sentences here are my translations from Jean Ziegler's French translations from the Dutch original; the second sentence is cited in de Castro, *Geography of Hunger,* 254. —Trans.

90 *Adam Hochschild points* . . . *hunger* See Hochschild, "Tug of War" (see note to p. 87).

11. A LIGHT IN THE DARKNESS: THE UNITED NATIONS

91 *One of the toughest* . . . *war* De Castro, *Geography of Hunger,* 257–58.

92 *"The case of France* . . . *machinery* Ibid., 266.

92 *Recovery was* . . . *painful* See Edgar Pisani's fine book *Le vieil homme et la terre* (Paris: Seuil, 2004) and, by the same author, *Vive la révolte!* (Paris: Seuil, 2006). [Born in 1918, Pisani is a French politician and agricultural advocate who, among many other positions, held the post of French minister of agriculture from 1961 to 1966. The two books noted here have unfortunately not been translated into English. —Trans.]

92 *As a result* . . . *diet* De Castro, *Geography of Hunger,* 267.

93 *In his State* . . . *fear* Links to both the full text and audio of the Four Freedoms speech are available on the FDR Library's website at docs.fdrlibrary.marist .edu/od4freed.html. —Trans.

93 *They* . . . *will endeavor* . . . *want* From the Atlantic Charter; an excellent, perfectly legible scan of the original Charter document itself is available at www.archives.gov/education/lessons/fdr-churchill/images/atlantic-char ter.gif. —Trans.

94 *When the fighting forces* . . . *world"* Sir John Boyd-Orr, "The Role of Food in Post-War Reconstruction," *International Labour Review* 47, no. 3 (1943): 279, available at labordoc.ilo.org/record/421022?ln=en. [This essay, originally delivered as an address in New York City in 1942 (the text of the speech is held in the Orr papers, box 1, folder 3, National Library of Scotland, Edinburgh), was also published in book form by the International Labour Office (ILO) in Geneva and Montreal and was widely read at the time.

Boyd-Orr was a Scottish physician, biologist, and politician who received the Nobel Peace Prize for his scientific research in nutrition and his work as the first director general of the FAO; he was made a baron in 1949. Boyd-Orr also contributed the foreword to the British and the first American editions of de Castro's *The Geography of Hunger.* —Trans.]

95 *We have come . . . creed* Franklin D. Roosevelt, State of the Union Message to Congress, January 11, 1944, available at www.presidency.ucsb.edu/ws /index.php?pid=16518#axzz1qwjjKTAc.

95 *Only fifty nations . . . 1945* Poland, which was not represented at the conference, soon signed the charter, to become the fifty-first original member state. See "History of the United Nations," www.un.org/en/aboutun /history. For the evolving membership of the UN, see "Member States: Growth in United Nations Membership, 1945–Present," www.un.org/en /members/growth.shtml. —Trans.

96 *ECOSOC* For more information on ECOSOC, see the organization's home page, www.un.org/en/ecosoc/index.shtml.

96 *Human Rights Council* For more information on the Human Rights Council, see the home page of the Office of the High Commissioner for Human Rights, www.ohchr.org/EN/Pages/WelcomePage.aspx.

12. JOSUÉ DE CASTRO, PHASE TWO: A VERY HEAVY COFFIN

98 *João Goulart, Leonel Brizola* Leonel Brizola was married to João Goulart's sister. Brizola was, like Goulart, a director of the PT and, on the eve of the coup d'état, governor of the state of Rio Grando do Sul and a federal deputy.

99 *the VAR-Palmares . . . member* Rousseff was arrested and tortured for several weeks by agents of the infamous Delegacia de Ordem Política e Social (DOPS, the Department of Political and Social Order) [which had been founded by the Vargas dictatorship in the 1930s and repurposed by the new military dictatorship. —Trans.]. She betrayed not one of her comrades. The VAR-Palmares (Vanguardia Armada Revolucionaria–Palmares, the Armed Revolutionary Vanguard–Palmares) was named after the legendary Palmares, in what is today the state of Alagoas, a federation of *quilombos* or hinterland settlements founded by rebellious fugitive slaves, which was effectively independent of the Brazilian government from about 1605 until its destruction in 1695. [For a survey of the history and historiography of the *quilombos*, see João Jose Flavio dos Santos and Reis Gomes, "Quilombo: Brazilian Maroons During Slavery," *Cultural Survival Quarterly* 35, no. 4 (Winter 2001), www.culturalsurvival.org/ourpublications/csq /article/quilombo-brazilian-maroons-during-slavery. —Trans.]

99 *In 1972, de Castro . . . environment* Paolo Freire, in his *Letters to Christina: Reflections on My Life and Work* (New York: Routledge, 1996), 238, quotes part of de Castro's address (citing in turn a special edition of *Polis*, "The Fight Against Hunger," 31):

> It is necessary to view the degradation of the economy of the underdeveloped countries as a contamination of their human environment caused by the economic abuses of the global economy; hunger, misery, high rates of illness, a minimum of hygiene, short average life expectancies—all of this is the product of the destructive

action of world exploitation according to the model of the capitalist economy. . . .

It is said that in the underdeveloped regions a concern for the qualitative aspects of life does not exist, only a concern for survival, that is, the battle against hunger, epidemic diseases, and ignorance. This attitude forgets that these are only the symptoms of a severe social illness: underdevelopment as a product of development. —Trans.

99 *Guararapes Airport in Recife*   The airport's full official name today is Recife/ Guararapes–Gilberto Freyre International Airport. —Trans.

100 *Everything leads . . . contradictory*   André Breton, Second Surrealist Manifesto, 1929. The second Manifesto is sometimes ascribed to Breton, and sometimes said to have been supervised by him but written by other members of the Surrealist movement. The text was first published in the twelfth and final issue of *La Révolution Surréaliste* (December 15, 1929), and in book form by both Kra and Gallimard in Paris the following year. This famous maxim has been translated in various ways; the version here is my own. —Trans.

100 The Masters and the Slaves   Gilberto Freyre, *The Masters and the Slaves: A Study in the Development of Brazilian Civilization*, trans. Samuel Putnam, 2nd ed. (New York: Knopf, 1956), a translation of *Casa grande e senzala: Formação da família brasileira sob o regime de economia patriarcal* (Rio de Janeiro: Maia and Schmidt, 1933). [This book was considered revolutionary in its day for proposing that the mixing of races in Brazilian society was a source of social and cultural strength. —Trans.]

100 *I encountered . . . Bastide*   See Jean Ziegler, *Les vivants et la mort* (Paris: Seuil, 1975; reissued in the series Points 1978, 2004). [This book has regrettably not yet been translated into English. —Trans.]

13. THE CRUSADERS OF NEOLIBERALISM

105 *Since the publication . . . corporations*   See Dan Morgan, *Merchants of Grain: The Power and Profits of the Five Giant Companies at the Center of the World's Food Supply* (New York: Viking, 1979). [Morgan's book is currently available in an Authors Guild Backinprint.com edition (Lincoln, Nebraska: iUniverse, 2000). —Trans.]

106 *Just ten corporations . . . beverages*   These figures are drawn from my report to the UNHRC titled *Promotion and Protection of All Human Rights, Civil, Political, Economic, Social and Cultural Rights, Including the Right to Development: Report of the Special Rapporteur on the Right to Food, Jean Ziegler*, A/HRC/7/5 (United Nations Human Rights Council, 2008;), available in six languages at www2.ohchr.org/english/bodies/hrcouncil/7session/reports.htm.

106 *"six companies control . . . bananas"*   Denis Horman, "Pouvoir et stratégie des multinationales de l'agroalimentaire," GRESEA (Groupe de Recherche pour une Stratégie Economique Alternative), April 2006, www.gresea.be /spip.php?article476. [Horman in turn credits for these figures the French edition of John Medeley's *Hungry for Trade: How the Poor Pay for Free Trade* (Halifax, NS: Fernwood, 2000). —Trans.]

106 *"From seeds to fertilizers . . . food"*   Doan Bui, *Les affameurs: Voyage au coeur de la planète faim* (Paris: Éditions Privé, 2009), 13. [Beauce is one of the most

productive agricultural regions in France, and Punjab a crucial grain-producing region in India. —Trans.]

106 *In his pioneering book . . . examined* See Gerald Gold, *Modern Commodity Futures Trading* (New York: Commodity Research Bureau, 1959). The CRB, founded in 1934, directs research in the production, distribution, consumption, and price movements of commodities and futures. [Gold's book was subsequently republished in many revised editions and remains widely available. —Trans.]

106 "Their purpose is . . . money" João Pedro Stedilé, "De la terre pour tous," in *Solutions locales pour un désordre global*, ed. Coline Serreau (Arles: Actes Sud, 2010), available at www.scribd.com/doc/48725253/Coline-Serreau -Solutions-Locales-Pour-Un-Desordre-Global-Decroissance-Ecologie -Capital-is-Me-BRF-Mondialisme-Latouche-Bourguignon. Stedilé is one of the principal directors of the Landless Rural Workers' Movement (MST) in Brazil. See also Jamil Chade, *O mundo não é plano: A tragédia silenciosa de 1 bilháo de famintos* (São Paulo: Editoras Saraiva e Virgília, 2010).

107 *Cargill is one . . . Farming* See *Cargill: A Threat to Food and Farming* (Washington, DC: Food & Water Watch, 2009), www.foodandwaterwatch .org/reports/cargill-a-threat-to-food-and-farming.

107 *As the report . . . fertilizers* Cargill spun off Mosaic in early 2011; see Michael J. de la Merced, "Cargill to Split Off Mosaic Unit in Complex Deal," *New York Times Dealbook*, January 18, 2011, dealbook.nytimes.com/2011/01/18 /cargill-to-spin-off-its-mosaic-unit-in-complex-deal. Merced writes, "The complicated tax-free transaction—worth more than $24 billion—will also help keep Cargill, one of the biggest American companies, private." According to Mosaic's home page, the company remains "the world's largest supplier of phosphate and potash"; see www.mosaicco.com. —Trans.

107 *Cargill has been . . . fresh* Food & Water Watch, *Cargill: A Threat to Food and Farming*, 6 (see note to p. 107).

108 *creates chemical byproducts . . . animals* Ibid., 6.

108 *By 2008, millions . . . crisis* Ibid., 8–9.

109 *In its Country Reports . . . beaten* See U.S. Department of State, Bureau of Democracy, Human Rights, and Labor, "2008 Human Rights Report: Uzbekistan," February 25, 2009, www.state.gov/j/drl/rls/hrrpt/2008 /sca/119143.htm. The report notes that children as young as nine have been observed working in the cotton harvest. The content and language of the Uzbekistan reports have hardly changed for years. The most recent report is available at www.state.gov/j/drl/rls/hrrpt/2010/sca/154489.htm. —Trans.

109 *speculative fever gripped . . . Continental* Morgan, *Merchants of Grain*, 210 (see note to p. 105). Since the French edition of this book was published, Cargill has reacquired Tradax and Continental has sold its trading division to Cargill. Oceangoing cargo ships, called "floats" in the jargon of the trade, generally carry cargoes of around 20,000 tons.

110 *Cargill also participated . . . crisis* Food & Water Watch, *Cargill: A Threat to Food and Farming*, 9 (see note to p. 107). [Cargill's website currently has links to two "financial services and commodity-trading" subsidiaries, Cargill Risk Management and Cargill Energy and Risk Management Solutions; for Cargill's own account of their activities, see www.cargill.com/products /financial-risk/index.jsp. —Trans.]

110 *Jim Prokopanko . . . chain"* See the interview with Prokopanko in Benjamin Beutler, "Konzentrierte Macht: Eine Handvoll transnationaler Konzerne kontrolliert weltweit Landwirtschaft und Nahrungsgütererzeugung," *Die Junge Welt*, November 23, 2009, Kapital und Arbeit, 9; available in German to subscribers at www.jungewelt.de/suche/index.php?and=Jim+Prokopan co&x=0&y=0&search=Suchen. [Prokopanko was at the time of the interview the president and chief executive officer of Mosaic, before it was spun off by Cargill—a position he still holds. He had previously risen through the corporate ranks at Cargill, becoming corporate vice president in 2004. See Prokopanko's profile in *Forbes* at people.forbes.com/profile/james-t-prokopanko/78965. —Trans.]

112 *According to the . . . combined* United Nations Development Programme, *Human Development Report 2010—20th Anniversary Edition: The Real Wealth of Nations—Pathways to Human Development*, hdr.undp.org/en/reports/global /hdr2010.

112 *As Sharad Pawar . . . alone* See the report on debt servitude in rural India released by the farmers' organization Ekta Parishad (New Delhi, 2011); Ekta Parishad points out the paradoxical nature of the gesture: the farmer kills himself with the very substance that is responsible for his intolerable indebtedness. [The report to which Jean Ziegler refers does not appear to be currently available on Ekta Prashad's website; however, its newsletters, reports, books, brochures, and resource materials are generally available at www.ektaparishad.com/media-section/publications. Ekta Prashad is a Gandhian national federation of more than nine hundred organizations advocating for land and forest rights for the poor, India's indigenous peoples, and dalits. —Trans.]

114 *In its rulings . . . law* Christophe Golay, *Droit à l'alimentation et accès à la justice* (doctoral thesis, Institut Universitaire des Hautes Études Internationales et du Développement, Geneva, 2009; Brussels: Bruylant, 2011).

114 *Food Corporation of India* For more information about the FCI, which is critically responsible for Indian food security, see fciweb.nic.in. —Trans.

115 *The anxiety of . . . hungry* Supreme Court of India, Civil Original Jurisdiction, Writ Petition (Civil) no. 196,2001, order of August 20, 2001; the decision is available for download from the Right to Food Campaign at www.rightto foodindia.org/orders/may203.html.

116 *"Any person . . . situation"* Golay, *Droit à l'alimentation* (see note to p. 114).

116 *In India, since . . . government* Colin Gonsalves, "Reflections on the Indian Experience," in *The Road to a Remedy: Current Issues in the Litigation of Economic, Social and Cultural Human Rights*, ed. John Squires, Malcolm Langford, and Bret Thiele (Sydney, NSW: Australian Human Rights Centre, Faculty of Law, University of New South Wales/Geneva: Centre for Housing Rights and Evictions, 2005), 177–82; the entire book is available for free download from the Global Initiative for Economic, Social and Cultural Rights website at 209.240.139.114/resources/esc-rights-library.

117 *indigenous forest-dwelling peoples* The "tribal peoples" of India, often known by the umbrella term Adivasi (a coinage dating to the 1930s with a connotation of former autonomy disrupted by colonialism), are a heterogeneous group of indigenous ethnic and tribal groups formally known as the Scheduled Tribes. There are 645 such nationally recognized groups, many

but not all living in the country's forests. Informally they are often called the *upajati* (clans, tribes, groups). —Trans.

117   *Colin Gonsalves . . . Brahmins*   Colin Gonsalves et al., *Right to Food*, vol. 1 (New Delhi: Human Rights Law Network, 2004). Before appealing to the federal supreme court, a plaintiff must have exhausted all local avenues of redress.

118   *And they won . . . . service*   See *High Court of South Africa (Witwatersrand Local Division), Lindiwe Mazibuko and Others v. The City of Johannesburg and Others,* Case No. 06/13885, judgment of April 30, 2008.

119   *"consensus of Washington"*   The term *consensus of Washington* refers to an array of informal agreements concluded throughout the 1980s and '90s among the principal Western multinational corporations, the Wall Street banks, the American Federal Reserve, the World Bank, and the IMF, all of which agreements aim to eliminate all financial regulating agencies, to liberalize markets, and to install "stateless global governance," or in other words, a unified and self-regulating global market. The theory behind the principles of this "consensus" was developed in 1989 by John Williamson, who was then chief economist and vice president of the World Bank.

119   *Plantu*   Plantu is the nom de plume of Jean Plantureux, who specializes in political satire. His work has appeared in the French newspaper *Le Monde* since 1972. —Trans.

14. THE HORSEMEN OF THE APOCALYPSE

120   *to a lesser extent, the World Bank*   In 2010, the International Finance Corporation (IFC), an affiliate of the World Bank, devoted $2.4 billion to subsistence agriculture in thirty-three countries in Africa, Asia, and Latin America. I borrow the term "horsemen of the Apocalypse" from one of my previous books, *Les nouveaux maîtres du monde et ceux qui leurs resistant* (Paris: Fayard, 2002; Points–Seuil, 2007). The IFC's website is www1.ifc.org.

120   *The World Bank . . . Robert Zoellnick*   Zoellnick directed the World Bank from 2006 to 2012. The recent change in leadership at the bank suggests that neoliberal orthodoxy there may soon wane at least somewhat. On March 23, 2012, U.S. president Barack Obama announced the nomination of Jim Yong Kim as the next president of the World Bank; Kim assumed office on July 1. The former president of Dartmouth College, Kim is the first World Bank president whose background is not in politics or finance, and the first to have prior experience in addressing health issues in developing countries, having served as chair of the Department of Global Health and Social Medicine at Harvard Medical School and as co-founder and executive director of Partners in Health, as well as in various positions at WHO (2003–6). Kim has announced his explicit commitment to "foster an institution that . . . prioritizes evidence-based solutions over ideology." See "Statement from Dr. Jim Yong Kim regarding his selection by the World Bank's Executive Directors as 12th President of the World Bank," press release no. 2012/398/SEC, April 16, 2012, web.worldbank.org/WBSITE /EXTERNAL/NEWS/0,,contentMDK:23170832~menuPK:3327604 ~pagePK:34370~piPK:34424~theSitePK:4607,00.html. —Trans.

121   *"The liberalization of . . . crisis"*   Marcel Mazoyer, address during the interactive panel discussion during the 47th executive session of UNCTAD, Food

Security in Africa: Lessons from the Recent Global Food Crisis, June 30, 2009. See also Marcel Mazoyer "Mondialisation libérale et pauvreté," *Alternative Sud* 4 (2003); and Marcel Mazoyer and Laurence Roudart, eds., *La fracture agricole et alimentaire mondiale: Nourrir l'humanité aujourd'hui et demain* (Paris: Universalis, 2005).

122 *Laurent Kabila . . . Maniema*   See my novel *L'Or de Maniema* (Paris: Seuil, 1996; Points–Seuil, 2011). [Not yet available in French. —Trans.]

123 *I was never . . . regard*   George Moose quit the diplomatic service with the arrival of the neoconservatives in the White House. [Although Moose resigned in 2001, he was promoted by President George W. Bush in 2002 to the rank of Career Ambassador, an honorary post. Since leaving the service he has taught a course at the George Washington University Elliott School of International Affairs, "Reinventing the United Nations," and led a study group on Africa as a fellow at the Harvard University Institute of Politics. —Trans.]

124 *One study . . . hunger*   Oxfam–Institute of Development Studies (IDS), *Liberalization and Poverty, Final Report to the Department for International Development* (London: DFID, 1999). See also *Rigged Rules and Double Standards: Trade, Globalization, and the Fight Against Poverty* (Oxford: Oxfam/Make Trade Fair, 2002), www.maketradefair.com/assets/english/report_english.pdf.

124 *Oxfam*   Oxfam was founded as the Oxford Committee for Famine Relief in 1942 [to campaign to send food supplies to starving civilians through the Allied naval blockade of Nazi-occupied Greece during World War II. Oxfam International today is a confederation of fifteen organizations working in more than ninety countries; Oxfam remains one of the world's leading NGOs in providing emergency food and drinking water, and fights hunger, poverty, and injustice worldwide. See www.oxfam.org. —Trans.]

124 *Of the fifty-three countries . . . economies*   Note that, depending upon one's political perspective, there are between fifty-three and fifty-seven countries in Africa. See Africa Check, "How Many Countries in Africa? How Hard Can the Question Be?" www.africacheck.org/reports/how-many-countries -in-africa-how-hard-can-the-question-be. —Trans.

125 *Haiti is today . . . tons*   Feyder, *Mordshunger*, 17ff (see note to p. 11).

125 *Since the 1990s . . . catastrophic*   Sally-Anne Way, *The Impact of Macroeconomic Policies on the Right to Food: The Case of Zambia* (London: Oxfam, 2001).

15. WHEN FREE TRADE KILLS

129 *aimed to relaunch . . . since*   In the WTO's own words, the Doha Round is a process of "trade negotiations among the WTO membership. Its aim is to achieve major reform of the international trading system through the introduction of lower trade barriers and revised trade rules" ("The Doha Round," www.wto.org/english/tratop_e/dda_e/dda_e.htm). —Trans.

129 *precooked cereal-and-pulse . . . biscuits*   For detailed descriptions of the special nutritional products distributed by the WFP, see www.wfp.org/nutrition /special-nutritional-products. —Trans.

130 *A woman widowed . . . liberalism?*   WFP memorandum, December 18, 2005. [Since the WFP does not offer the texts of all its memoranda on its website, I

have been unable to locate the official English-language version of this document. The translation here is my own, from Jean Ziegler's French. I presume that the memorandum referred to was originally released in English. —Trans.]

130 *In 1943 . . . fronts* See Jean Drèze, Amartya Sen, and Athar Hussain, eds., *The Political Economy of Hunger: Selected Essays* (Oxford: Clarendon, 1995).

16. SAVONAROLA ON LAKE GENEVA

134 *"I am neither . . . signed"* Sonia Arnal, "Pascal Lamy: 'Je ne suis ni optimiste ni pessimiste. Je suis activiste,'" *Le Matin Dimanche*, Lausanne, February 12, 2011, new.lematin.ch/actu/economie/pascal-lamy-%C2%ABje-ne-suis-ni-optimiste-ni-pessimiste-je-suis-activiste%C2%BB-89076.

135 *One of his books . . . markets* See Pascal Lamy, *L'Europe en première ligne*, with a preface by Éric Orsenn (Paris: Seuil, 2002), particularly the chapter titled "Les cent heures de Doha," 147ff.

135 *Jean-François Noblet* For more information on the work of the ecologist Jean-François Noblet, see his website, noblet.me. —Trans.

136 *Olivier De Schutter* De Schutter, a professor of international human rights law, succeeded Jean Ziegler as UN Special Rapporteur on the Right to Food in 2008. —Trans.

137 *"We agree to ensure . . . manner"* WTO Doha Work Programme Ministerial Declaration (adopted on December 18, 2005), para. 6, www.wto.org/english/thewto_e/minist_e/min05_e/final_text_e.htm.

137 *The report argues . . . farmers* *Background Document Prepared by the UN Special Rapporteur on the Right to Food, Mr. Olivier De Schutter, on His Mission to the World Trade Organization (WTO), Presented to the Human Rights Council in March 2009* (background study to UN doc. A/HRC/10/005/Add.2), January 5, 2009, available at www.srfood.org/images/stories/pdf/otherdocuments/9-srrtfre portmissionwto-1-09.pdf.

138 *Virtually all NGOs . . . Doha Round* See in particular the "Note conceptuelle pour le Forum social mondial (FSM) 2011," written by the World Social Forum's scientific committee, headed by Samir Amin, in preparation for the Forum's conference in Dakar, February 6–11, 2011, fsm2011.org/fr/note-conceptuelle [in French only —Trans.]. See also the document presented by La Via Campesina and adopted by the Forum's plenary assembly, "Dakar Appeal Against the Land Grab," www.viacampesina.org/en/index.php?option=com_content&view=article&id=1040:dakar-appeal-against-the-land-grab&catid=23:agrarian-reform&Itemid=36; and, also from La Via Campesina, "Final Declaration of the Social Movements Assembly WSF 2011, February 10th, Dakar (Senegal)," www.viacampesina.org/en/index.php?option=com_content&view=article&id=1034:final-declaration-of-the-social-movements-assembly-wsf-2011-february-10th-dakar-senegal&catid=25:world-social-forum&Itemid=34; and, from the Forum, "Declaration by the Convergence Assembly on Social and Economy and Fair Trade," fsm2011.org/en/social-and-economy-and-fair-trade.

138 *"The lord . . . mind"* Ecclesiastes 28:28, Revised Standard Version.

17. A BILLIONAIRE'S FEAR

142 *The policies governing . . . FAO* See the WFP's Mission Statement, which is prefaced: "In December 1994, WFP's governing body adopted the WFP Mission Statement, the first for an United Nations organization," www.wfp .org/about/mission-statement. —Trans.

143 *Food for Work* For more information on the WFP's Food for Assets program, see www.wfp.org/food-assets. —Trans.

144 *The two autonomous regions* Following the 2008 South Ossetia war, in which Russia intervened to defend the two breakaway regions and Georgia was decisively defeated, the two self-declared republics reaffirmed their independence, which, however, has been recognized by only the Russian Federation and a handful of other UN member states (Nicaragua; Venezuela; and the tiny Pacific island states of Nauru, Tuvalu, and Vanuatu). The vast majority of the world's countries and international organizations continue to consider the two regions to be parts of Georgia. —Trans.

148 *"Dear Mr. Ziegler. . . succeed"* Since the original letters were not available to me, the texts here are my translations from Jean Ziegler's French renderings of Morris's letters (which I presume were written in English). —Trans.

150 *Exhausted, worn out . . . 2007* Since leaving the WFP, James T. Morris has served as an independent director of Old National Bancorp in Evansville, Indiana, and as president of the company that owns the Indiana Pacers men's basketball team.

18. VICTORY OF THE PREDATORS

152 *Before 2009 . . . Territories* For the WFP's 2009 and 2010 statistics on its school meal programs, see docustore.wfp.org/stellent/groups/public /documents/newsroom/wfp236983.pdf. —Trans.

154 *It would be . . . WFP* See "Food Aid for Africa: When Feeding the Hungry Is Political—A United Nations Agency Under Attack," *The Economist*, March 18, 2010, www.economist.com/node/15731546.

20. JALIL JILANI AND HER CHILDREN

160 *Waliur Rahman* By coincidence, a BBC reporter of the same name (which is relatively common in Bangladesh) filed a story in 2005 naming the country the world's most corrupt, tied with Chad. See Waliur Rahman, "Bangladesh Tops Most Corrupt List" BBC News, October 18, 2005, news .bbc.co.uk/2/hi/south_asia/4353334.stm. —Trans.

160 *Gulshan* Ironically, Gulshan is also the name of the wealthiest district of Dhaka. It is obviously important not to confuse the two. —Trans.

161 *All the peoples . . . majority* Bangladesh is a very homogeneous country, with 98 percent of the population being ethnically Bengali and nearly 90 percent Muslim. The country's approximately one hundred tribal groups include Hindus, Buddhists, Christians, and animists and are mainly of Sino-Tibetan origin. —Trans.

162 *The legal minimum wage . . . $11.00* See "Spectrum Disaster—New Info and Demands," Clean Clothes Campaign, April 24, 2005, www.cleanclothes

.org/news/spectrum-update-new-info-a-demands. The fluctuating ex-change rate for Bangladesh's currency in this period had a considerable effect on the value of wages for the country's textile workers. From 2004 to mid-2006 the Bangladeshi taka decreased steadily in value from about 60 to 70 takas to the U.S. dollar; the average in 2005, when Jean Ziegler met Jalil Jilani, was 63.75. The currency traded steadily just below 70 to the dollar until late 2010, when it began to fall sharply in value; by April 2012, the taka was trading at about 82 to the dollar. Bangladesh raised its minimum wage for garment workers in 2006 and again in 2010, to 3,000 takas per month; see "Bangladesh Increases Garment Workers' Minimum Wage," BBC News, July 23, 2010, www.bbc.co.uk/news/world-south-asia-10779270. Nonetheless, according to the International Labour Organisation's *Global Wage Report 2010/2011*, Bangladesh still has the lowest minimum wage in Asia; see Iminul Islam, "Minimum Wage of Bangladesh Lowest in Asia," *Daily Newspaper*, December 17, 2010, available at www.cawinfo.org/2011/01 /minimum-wage-of-bangladesh-lowest-in-asia. —Trans.

162 *Clean Clothes Campaign*   See the Clean Clothes Campaign, "About Us," www .cleanclothes.org/about-us.

162 *The CCC has calculated . . . 33 cents*   Berne Declaration bulletin, 2005.

162 *On the night . . . rubble*   See "Factory Collapsed—Bangladeshi Workers Buried Alive," Clean Clothes Campaign, April 1 [*sic*; should be April 10 or 11 —Trans.], 2005, www.cleanclothes.org/news/Bangladeshi-garment -workers-buried-alive. Sixty-four workers died and seventy-four were in-jured, including many left permanently handicapped. For a series of highly detailed reports on the Spectrum Sweater disaster and its aftermath, in-cluding the attempts to secure justice and compensation on behalf of the victims' families, see the Clean Clothes Campaign, www.cleanclothes.org /search?searchword=spectrum&ordering=&searchphrase=all. —Trans.

21. THE DEFEAT OF JACQUES DIOUF

166 The Lords of Poverty   Graham Hancock, *The Lords of Poverty: The Power, Prestige, and Corruption of the International Aid Business* (London: Macmillan, 1989).

166 The Ecologist *is . . . hunger*   *The Ecologist* 21, no. 2, *FAO Special Issue: The UN Food and Agriculture Organization: Promoting World Hunger* (March/ April 1991); available at exacteditions.theecologist.org/exact/browse /307/308/5643/2/1?dps=on. [The issue opens with an open letter to Edouard Saouma, then director general of the FAO, written by Nicholas Hildyard and endorsed by leading NGOs from all over the world: "It is your policies that are at fault, not peasants or lack of finance. Whether in agriculture, in forestry, or in aquaculture, you have promoted policies which benefit the rich and powerful at the expense of the livelihoods of the poor. Policies that are, in effect, systematically creating the conditions for mass starvation" (43). —Trans.]

166 *As for the BBC . . . money*   "Food Summit 'Waste of Time,'" BBC News World Edition, June 13, 2002, news.bbc.co.uk/2/hi/2042664.stm. [This article was published immediately following the FAO's second World Food Summit. —Trans.]

167 *By comparison . . . subsidies*   See *The Director-General's Medium Term Plan 2012–13 (Reviewed) and Programme of Work and Budget 2012–13* (Rome: FAO, 2011), www.fao.org/docrep/meeting/021/ma061e.pdf. —Trans.

167 *The FAO owes . . . 2011*   [As of January 1, 2012,] Diouf was replaced by José Graziano da Silva, a warm-hearted [American-born] Brazilian agronomist. Graziano is well qualified: he served in the first Lula cabinet as Extraordinary Minister for Food Security from 2003 to 2004 and was responsible for implementing the Fome Zero program [which lifted 28 million people above the national poverty line during Lula's eight years in office —Trans.].

171 *On paper . . . population*   See Hans Christof von Sponeck, *A Different Kind of War: The UN Sanctions Regime in Iraq*, with a foreword by Celso N. Amorim (London: Berghahn Books, 2006), a translation of *Ein anderer Krieg: Das Sanktionsregime der UNO im Irak* (Hamburg: Hamburger Edition, 2005).

171 *Thus, gradually . . . medicine*   See Hans Christof von Sponeck and Andreas Zumach, *Irak, Chronik eines gewollten Krieges: Wie die Weltöffentlichkeit manipuliert und das Völkerrecht gebrochen wird* (Cologne: Kiepenheuer und Witsch, 2003). [This work, whose title means "Iraq, Chronicle of an Unprovoked War: How World Public Opinion Is Manipulated and International Law Broken," has not been translated into English. —Trans.]

173 *"Even if not . . . war"*   *Report of the Second Panel Established Pursuant to the Note by the President of the Security Council of 30 January 1999 (S/1999/100), Concerning the Current Humanitarian Situation in Iraq* (Annex II of S/1999/356, March 30, 1999), available at www.casi.org.uk/info/panelrep.html.

173 *"How ironic it is . . . destruction"*   Hans von Sponeck, "After the Journey—A UN Man's Open Letter to Tony Blair," *New Statesman*, September 23, 2010, www.newstatesman.com/middle-east/2010/09/iraq-humanitarian-sanctions.

174 *"We have heard . . . worth it"*   The most complete transcript available online of Stahl's interview with Albright is included in Douglas E. Hill, "Albright's Blunder," *Irvine Review* (2002), available at web.archive.org/web/20030603215848/ http://www.irvinereview.org/guest1.htm. Hill, however, is very supportive of Albright and highly critical of Stahl. —Trans.

175 *His predecessor . . . uproar*   Halliday resigned to protest the sanctions regime, saying publicly, like Marc Bossuyt, that it amounted to genocide. —Trans.

175 *"This man . . . speak"*   James Rubin, Spokesman, U.S. Department of State, daily press briefing, quoted in von Sponeck, *Different Kind of War*, 4 (see note to p. 171).

175 *"As the UN Humanitarian Coordinator . . . propaganda"*   Von Sponeck, *Different Kind of War*, 211.

176 *The American bombing . . . program*   Certain accounts from the oil-for-food program were transferred to the Iraqi Development Fund administered by the American proconsul in Baghdad, Paul E. Bremer. See Djacoba Liva Tehindrazanarivelo, *Les sanctions des Nations unies et leurs effets secondaires* (Paris: PUF, 2005).

22. A GREAT LIE

180 *More than 600 million . . . 2011*   For statistics through 2011, see Early Warning: Risks to Global Civilization, "Global Biofuel Production," earlywarn.blogspot.com/2011/03/global-biofuel-production.html. For slightly

different figures, see "Biofuels" on the BP website, www.bp.com/section
genericarticle800.do?categoryId=9037217&contentId=7068633. —Trans.

180 *world production . . . 2011* See Benoît Boisleux, *Impacts des biocarburants sur
l'équilibre fondamental des matières premières aux États-Unis* (Zurich, 2011).

180 *The dry regions . . . land* See Robin P. White and Janet Nackoney, *Drylands,
People, and Ecosystem Goods and Services: A Web-Based Geospatial Analysis*
(Washington: World Resources Institute, 2003), www.wri.org/publication
/drylands-people-and-ecosystem-goods-and-services.

181 *The destruction of ecosystems . . . pastoralists* On the causes of ecosystem destruction in Europe, see Serreau, *Solutions locales* (see note to p. 106); see
also the excellent film of the same title (information is available at www
.solutionslocales-lefilm.com).

182 *According, again . . . facilities* See Riccardo Petrella, *Le manifeste de l'eau: Pour
un contrat mondial*, preface by Mario Soares, 2nd expanded ed. (Lausanne:
Page Deux, 1999); [available in English as *The Water Manifesto: Arguments for
a World Water Contract* (London: Zed, 2001). Petrella has recently updated
his book for the new century: see *Le manifeste de l'eau pour le XXIe siècle: Pour
un pacte social de l'eau* (Anjou, Québec: Fides, 2008) —Trans.]. See also Guy
Le Moigne and Pierre-Frédéric Ténière-Buchot, "Les grands enjeux liés à
la maîtrise de l'eau," *De l'eau pour demain*, special issue of *Revue Française de
Géoéconomie*, no. 4 (Winter 1997/98): 37–46.

183 *Noël Mamère* Mamère is a French TV entertainer turned politician and
ecological activist, and a member of the French Green Party. —Trans.

183 *Peter Brabeck-Letmathe . . . Nestlé* Peter Brabeck Letmathe, in an interview
with the Sunday *Neue Zürcher Zeitung*, "'Dann gibt's nichts mehr zu essen':
Nestlé-Chef Peter Brabeck warnt im Interview vor Agro-Treibstoffen," *Neue
Zürcher Zeitung*, March 23, 2008, available in German only at www.nzz.ch
/nachrichten/wirtschaft/aktuell/dann_gibts_nichts_mehr_zu_essen_1
.693850.html.

183 *In addition . . . biofuel* See "Biofuel Policies in OECD Countries Costly
and Ineffective, Says Report," www.oecd.org/document/28/0,334
3,fr_2649_33717_41013916_1_1_1_1,00.html; the report itself, the
OECD's *Economic Assessment of Biofuel Support Policies*, and a PowerPoint
presentation of its conclusions, may be downloaded from the same
webpage.

183 *"a large-scale effort . . . reduce it"* "Priced Out of the Market," editorial,
New York Times, March 7, 2008, www.nytimes.com/2008/03/03/opinion
/03mon1.html.

23. BARACK OBAMA'S OBSESSION

184 *In his State of the Union . . . priority* The complete text of President Obama's
2011 State of the Union address is available from the *Huffington Post* at
www.huffingtonpost.com/2011/01/25/obama-state-of-the-union-_1_n
_813478.html; the White House's own coverage is at www.whitehouse.gov
/state-of-the-union-2011. Both sites feature complete video of the speech.
—Trans.

186 *biofuels . . . stomachs* Among countless other instances, see David Zilberman

et al., "The Economics of Biofuel," 24, a presentation to the FAO by a team from Berkeley's program in agricultural and resource economics, www.fao.org/fileadmin/user_upload/foodclimate/presentations/EM56 /Zilberman.pdf. Jean Ziegler cites a Swiss edition (in French) of *Amnesty International* magazine, Bern, September 2008. —Trans.

24. THE CURSE OF SUGARCANE

187 engenhos *(sugarcane plantations)* In Brazil in the colonial era, an *engenho* encompassed a sugar plantation's entire property: the cane fields, the buildings where cane was processed (the *casa de engenho*), the landowner's residence (the *casa grande*), and the slaves' quarters *(senzala)*.

189 *The implementation of . . . farms* Michel Duquette, "Une décennie de grands projets: Les leçons de la politique énergétique du Brésil," *Tiers-Monde* 30, no. 120 (1989): 907–25.

190 *One FAO expert . . . plantations* Ricardo Abramovay, *Policies, Institutions and Markets Shaping Biofuel Expansion: The Case of Ethanol and Biodiesel in Brazil* (Rome: FAO, 2009), 10.

190 *The monopolization exacerbates . . . production* See Frances Moore Lappé and Joseph Collins, *Food First: Beyond the Myth of Scarcity* (New York: Houghton Mifflin, 1977; New York: Ballantine, 1979). [Jean Ziegler cites specifically a French edition of this book, *L'industrie de la faim: Par-delà le mythe de la pénurie* (Montreal: L'Étincelle, 1977), 213. —Trans.]

190 *As for rural families . . . discrimination* One FAO study specifically examines discrimination against single women heads of rural households and the ways that they have been affected by the implementation of the Pro-alcohol plan; see Andrea Rossi and Yianna Lambrou, *Gender and Equity Issues Gender and Equity Issues in Liquid Biofuels Production: Minimizing the Risks to Maximize the Opportunities* (Rome: FAO, 2008), www.fao.org/docrep/010 /ai503e/ai503e00.htm. For a summary of the report, see "Large-Scale Biofuel Production May Increase Marginalization of Women: New Study on Biofuel Production Focuses on Gender," FAO, Rome, April 21, 2008, www.fao.org/newsroom/en/news/2008/1000830/index.html. —Trans.

191 *Ethical Sugar* For Ethical Sugar's overall view of the Brazilian biofuel industry, see Ben Richardson, Markku Lehtonen, and Siobhán McGrath, *An Exclusive Engine of Growth: The Development Model of Brazilian Sugarcane* (Ethical Sugar, 2009); available at www.sucre-ethique.org/IMG/pdf/Ethical _Dugar_social_report_2009_-_Brazil.pdf. —Trans.

191 *Paulo Vanucci* Vanucci was the Brazilian minister of human rights from 2005 to 2010. —Trans.

192 *According to an estimate . . . 2050* See World Bank, "Assessment of the Risk of Amazon Dieback: Main Report" (Climate Change and Clean Energy Initiative, Environmentally and Socially Sustainable Development Department, Latin America and Caribbean Region), February 4, 2010, 58, www.bicusa.org/en/Document.101982.aspx. See also Britaldo Silveira Soares-Filho et al., "Modelling Conservation in the Amazon Basin," *Nature* 440 (March 23, 2006): 520–23, available to subscribers or for purchase at www.nature.com/nature/journal/v440/n7083/full/nature04389.html. —Trans.

192 *As David and Marcia . . . families* David Pimentel and Marcia H. Pimentel, *Food, Energy, and Society,* 3rd ed. (Boca Raton, FL: CRC Press, 2007), 294.

193 *Many live and suffer . . . herd* Clemens Höges, "Les esclaves brésiliens de l'éthanol: Derrière le miracle des agrocarburants," *Le Courrier International,* April 30, 2009, www.courrierinternational.com/article/2009/04/30 /les-esclaves-bresiliens-de-l-ethanol. [This article was originally published in *Der Spiegel* in German and is available in a slightly different version in English at archive.truthout.org/012409C. —Trans.]

194 *"due to existing . . . risks"* "Large-Scale Biofuel Production May Increase Marginalization of Women" (see note to p. 190). [See also Rossi and Lambrou, *Gender and Equity* (see note to p. 190). And see also "The Situation of Women in Rural Areas," in *Women, Migration, Environment and Rural Development Policy in Brazil* (Rome: FAO, Economic and Social Development Department, no date), www.fao.org/DOCREP/x0210e/x0210e03.htm; and "The Double Work Burden: Work at Home and in the Field," FAO, Economic and Social Development Department, no date, www.fao.org /DOCREP/x0210e/x0210e05.htm. —Trans.]

194 *In 2004 . . . plantations* See Edward Smeets et al., *Sustainability of Brazilian Bio-ethanol* (Utrecht: University of Utrecht Copernicus Institute, Department of Science, Technology, and Society, August 2006), available at www.biofuels -platform.ch/en/media/download.php?get=195. The source of these statistics is Simon Schwartzman and Felipe Farah Schwartzman, "O trabalho infantil no Brasil," *Jornal do Brasil* (2004), available at www.schwartzman.org.br /simon/pdf/trab_inf2004.pdf. See also *The Good Practices of Labour Inspection in Brazil: The Eradication of Labour Analogous to Slavery* (Geneva: International Labor Organization, 2010), www.ilo.org/wcmsp5/groups/public/---ed _norm/---declaration/documents/publication/wcms_155946.pdf.

194 *Gilberto Freyre's famous book* See note to p. 100.

25. CRIMINAL RECOLONIZATION

196 *FoodFirst Information and Action Network* "FIAN's vision is a world free from hunger, in which every woman, man and child can fully enjoy their human rights in dignity, particularly the right to adequate food, as laid down in the Universal Declaration of Human Rights and other international human rights instruments" (www.fian.org/about-us/introduction).

196 *Centre Europe–Tiers Monde* For information on this organization, see www .cetim.ch/en/index.php?currentyear=&pid=. —Trans.

196 *"First they took . . . Africa"* Presumably Montwedi made his remarks in English; I have had to translate them from Jean Ziegler's French text here. —Trans.

196 *Colombia provides . . . example* I rely here upon reports from Human Rights Watch and Amnesty International. See especially an issue of Amnesty International Switzerland's magazine titled *Agrocarburants: Réservoirs pleins et ventres vides, Magazine Amnesty* 54 (September 2008), and in particular the article "Du sang sous les palmiers colombiens," www.amnesty.ch/fr /actuel/magazine/2008-3/monoculture-palme-africaine. [See also "Voix dissidentes sous pression," www.amnesty.ch/fr/actuel/magazine/2008-3

/journaliste-colombie-temoin-indesirable; "Réservoirs pleins et ventres vides," www.amnesty.ch/fr/actuel/magazine/2008-3/reservoirs-pleins-et -ventres-vides;"Lesfemmeslaisséespourcompte,"www.amnesty.ch/fr/pays /ameriques/colombie/docs/2011/rapport-violence-sexuelle; "Des engagements sociaux multiples," www.amnesty.ch/fr/actuel/magazine/2008-3 /colombie-salazar. Unfortunately, none of these articles appear to have been translated into English. —Trans.]

197 *Between 2002 and 2007 . . . paramilitaries*   See "Bilan noir des droits humains en Colombie: Le président colombien Alvaro Uribe s'en prend à la Cour supreme," *Le Temps*, Geneva, September 20, 2008, www.letemps.ch /QueryResult?offset=120 (available in French to subscribers only).

197 *In 1993 . . . peace*   According to an exclusive report written for Sustainable Security, "The Colombian government has recently begun a process of returning land to the inhabitants of the river basins of Curvaradó and Jiguamiandó and reparation for victims of violence. The move is encouraging, but it might not be enough to solve the problems. The history of violence can repeat itself any moment, as long as the causes that led to the banishment and violence are not addressed and those responsible are not punished." See Amira Armenta, "Conflict, Poverty and Marginalisation: The Case of Curvaradó and Jiguamiandó (Urabá, Colombia)," Sustainable Security, July 2011, sustainablesecurity.org/article/conflict-poverty-and -marginalisation-case-curvarad%C3%B3-and-jiguamiand%C3%B3 -urab%C3%A1-colombia. —Trans.

197 *Human rights organizations . . . plantations*   According to Amnesty International; see note to p. 196.

197 *the long battle . . . Ferrari*   See Sergio Ferrari, "Colombie: Une communauté paysanne tente de récupérer ses terres—Le label 'Bio suisse' impliqué dans une affaire d'expropriation?" *Le Courrier*, Geneva, March 15, 2011, available at www.kipa-apic.ch/index.php?pw=&na=0,0,0,0,f&ki=218084. [A good summary in English of the Las Pavas case, including recent developments, is *Report, Independent Commission Land Conflict—Las Pavas–Bolívar, Colombia*, Body Shop–Christian Aid, June 2010, www.thebodyshop.com/_en /_ww/services/pdfs/AboutUs/LasPavasReview.pdf. —Trans.]

198 *Consider what is happening . . . Africa*   See *Africa Up for Grabs: The Scale and Impact of Land Grabbing for Agrofuels*, Friends of the Earth Europe and Friends of the Earth Africa, June 2010, www.foei.org/en/resources/publications /pdfs/2010/africa-up-for-grabs/view.

200 Jatropha curcas . . . *oil*   *Jatropha curcas* is a flowering, poisonous, semi-evergreen shrub native to the American tropics and highly resistant to arid conditions, allowing it to be grown in desertlike climates; its seeds or nuts can be processed to yield a high-quality biodiesel fuel usable in standard diesel engines. —Trans.

200 *In 2008 . . . contract*   The *Financial Times* published a series of detailed articles reporting on the secret agreement and its unraveling. See especially Tom Burgis, Song Jung-a, and Christian Oliver, "Daewoo to Cultivate Madagascar Land for Free," *Financial Times*, November 19, 2008, www .ft.com/intl/cms/s/0/6e894c6a-b65c-11dd-89dd-0000779fd18c.html. See also, by the same reporters, "Daewoo to Pay Nothing for Vast Land

Acquisition," *Financial Times,* November 20, 2008, www.ft.com/intl/
cms/s/0/b0099666-b6a4-11dd-89dd-0000779fd18c.html; Javier Blas,
"S[outh] Koreans to Lease Farmland in Madagascar," *Financial Times,*
November 19, 2008, www.ft.com/intl/cms/s/0/ea8de830-b5d9-11dd-ab71
-0000779fd18c.html; Javier Blas and Tom Burgis, "Madagascar Scraps
Daewoo Farm Deal," *Financial Times,* March 18, 2009, www.ft.com/intl
/cms/s/0/7e133310-13ba-11de-9e32-0000779fd2ac.html; and Javier Blas,
"UN Moves to Curb Farmland Grabs," *Financial Times,* March 25, 2012,
www.ft.com/intl/cms/s/0/083aab3a-7697-11e1-8e1b-00144feab49a.html.
—Trans.

201 *Sierra Leone . . . world* See United Nations Development Program, *Human
Development Report 2010, 20th Anniversary Edition—The Real Wealth of Nations:
Pathways to Human Development* (New York: Palgrave Macmillan, 2010), avail-
able at hdr.undp.org/en/media/HDR_2010_EN_Complete_reprint.pdf.

201 *Spread out among . . . highlands* Joan Baxter, "Le cas Addax Bioenergy," *Le
Monde Diplomatique,* January 2010, www.monde-diplomatique.fr/2010/01
/BAXTER/18712.

203 *an independent field study . . . false* Coastal and Environmental Services,
Environmental Social and Health Impact Assessment (ESHIA), undertaken
on behalf of Addax Bioenergy by Coastal and Environmental Services
for the Makeni Sugarcane to Ethanol Biofuel Project in Sierra Leone
(October 2009). [This report is supposed to be available at the Coastal and
Environmental Services Public Documents webpage, though as of May
2012 it was not: www.cesnet.co.za/publicdocs.html. The report was rated
as exemplary in an independent study commissioned by the European
Commission to assess the value of biofuel environmental impact assess-
ments (EIAs) in determining a project's sustainability and compliance
with the European Union Renewable Energy Directive (EU RED). See
Oskar Englund et al., *Environmental Impact Assessments: Suitable for Supporting
Assessments of Biofuel Sustainability?—Analysis of EIAs from the Perspective of EU
Sustainability Requirements for Biofuels* (Technical Report for the EU Biofuel
Baseline Project) (Gothenburg: Department of Energy and Environment,
Division of Physical Resource Theory, Chalmers University of Sociology,
2011), publications.lib.chalmers.se/records/fulltext/local_146738.pdf. For
Addax's perspective, see also *Q&A: Addax Bioenergy Sugarcane Ethanol Project
in Makeni, Sierra Leone,* www.addax-oryx.com/AddaxBioenergy/Addax
-Bioenergy-FAQ.pdf. —Trans.]

204 *The Canadian Renewable Fuels Association . . . "absurd"* Lauren Etter, "UN
Is Urged to Disavow 'Rogue' Biofuels Remarks," *Wall Street Journal,*
November 13, 2007, A6. [This article is not available online. —Trans.]

26. THE "TIGER SHARKS"

207 *With its powerful jaws* Jean Ziegler could have added here two more facts
about tiger sharks that speak very much to his metaphor. After the great
white shark, the tiger is responsible for more confirmed attacks on humans
than any other species. The tiger also seems to be one of the most voracious of all shark species, often swallowing its prey whole; sometimes nick-
named the "ocean's garbage can," tigers have been found with all manner

of inorganic objects in their gut, including license plates, oil cans, tires, fishing gear, boat furniture, and baseballs. Unlike commodities market speculators, however, the tiger shark is an endangered species. —Trans.

208 *Société Générale* Société Générale S.A. is one of the oldest banks in France and a globally active financial services company. It is the second-largest French bank behind BNP Paribas and the eighth largest bank in the European zone. —Trans.

208 *Jérôme Kerviel . . . bank* In October 2010, the Eleventh Chambre Correctionnelle (Criminal Court) of Paris sentenced Jérôme Kerviel to five years in prison (with the possibility of release after three years) and a fine for damages and interest of 4.9 billion euros. [Kerviel's appeal of his conviction is currently pending. —Trans.]

208 *By contrast . . . 51.85 percent* GAIA Capital Advisors, GAIA World Agri Food Fund February 2011 report, www.gaiacap.ch/newsletters/gwa_nav _feb_11.pdf company. (In Greek mythology, Gaia was the Earth personified as a goddess.)

208 *the purchase (or sale) . . . markets* Nicholas Kaldor, "Speculation and Economic Stability," *Review of Economic Studies* 7, no. 1 (October, 1939): 1–27, at 1.

208 *"Speculation is . . . reward"* Miguel Robles, Maxime Torero, and Joachim von Braun, *When Speculation Matters*, IFPRI Issue Brief 57 (Washington, DC: International Food Policy Research Institute, 2009), 2, available at www.google.com/url?sa=t&rct=j&q=&esrc=s&source=web&cd=1&ved =0CCIQFjAA&url=http%3A%2F%2Fwww.abhatoo.net.ma%2Findex .php%2Ffre%2Fcontent%2Fdownload%2F13502%2F230576%2Ffile %2FWhen_Speculation_matters.pdf&ei=9TijT8w7ht3RAfjogSc&usg =AFQjCNGqVv95fz4h_oILBRzom_-q96hN-Q&sig2=8YnvbjPdPu1N _r6beJpVkg.

209 *the first derivatives . . . 43 percent* Olivier Pastré, "La crise alimentaire mondiale n'est pas une fatalité," in *Les nouveaux équilibres agroalimentaires mondiaux*, ed. Pierre Jacquet and Jean-Hervé Lorenzi (Paris: Presses Universitaires de France–PUF, in the series *Les Cahiers du Cercle des Économistes*, 2011), 29.

209 *this market is . . . amplifies* Ibid.

211 *Many factors are . . . speculation* See Jacquet and Lorenzi, *Les nouveaux équilibres agroalimentaires mondiaux* (see note to p. 209).

211 *The crisis of 2008 . . . 2006* This idea is defended particularly by Philippe Chalmin, in *Le monde a faim: Quelques réflexions sur l'avenir agricole et alimentaire de l'humanité au XXIe siècle* (Paris: Bourin Éditeur, 2009). El Niño is a seasonal warm current in the Pacific originating off the coast of Peru near the equator, which has in the last several years caused numerous extreme weather events.

211 *In 2008 . . . 2006* See the FAO's *The State of Agricultural Commodity Markets 2009: High Food Prices and the Food Crisis—Experiences and Lessons Learned* (Rome, 2009), www.fao.org/docrep/012/i0854e/i0854e00.htm.

212 *in March [2008] . . . ton* See Chalmin, *Le monde a faim* (see note to p. 211).

212 *As for corn . . . market* Ibid.

213 *"That speculation on . . . financiers"* Ibid., 45.

213 *The hedge funds . . . 2007* Laetitia Clavreul, "La spéculation sur les matières premières affole le monde agricole," *Le Monde*, April 24, 2008, available to subscribers or for purchase at www.lemonde.fr/cgi-bin/ACHATS

/acheter.cgi?offre=ARCHIVES&type_item=ART_ARCH_30J&objet
_id=1033563&xtmc=laetitia_clavreul&xtcr=4.

213 *The Senate denounced . . . applied"* See U.S. Senate Committee on Homeland
Security and Governmental Affairs, "Investigations Subcommittee Releases
Levin–Coburn Report on Excessive Speculation in the Wheat Market:
Report Calls for Clampdown on Index Traders Buying Wheat Futures,"
June 23, 2009, www.hsgac.senate.gov/media/majority-media/investigations
-subcommittee-releases-levin-coburn-report-on-excessive-speculation-in
-the-wheat-market. The Levin–Coburn Report itself, released by the Senate
Permanent Committee on Investigations in advance of hearings in July 2009,
is not currently available online, but its conclusions are summarized in this
press release. Video recordings and the print transcripts of the hearings are
available at www.hsgac.senate.gov/subcommittees/investigations/hearings
/excessive-speculation-in-the-wheat-market. Jean Ziegler refers to Paul-
Florent Montfort, "Le Sénat américain dénonce la spéculation excessive sur
les marches à terme agricoles: Rapport du sous-comité permanent du Sénat
des États-Unis en charge des enquêtes," Mouvement pour une Organization
Mondiale de l'Agriculture (2009), www.momagri.org/FR/articles/Le-Senat
-americain-denonce-la-speculation-excessive-sur-les-marches-a-terme
-agricoles_538.html. This article provides more details from the report than
the Senate's press release does, albeit in French. —Trans.

214 *"Many producer countries . . . so on)"* See Clavreul, "La speculation" (see note
to p. 213).

215 *West African CFA francs* The CFA franc is the common currency of the
eight members of the UEMOA (Union Économique et Monétaire Ouest
Africaine, the West African Economic and Monetary Union): Benin,
Burkina Faso, Ivory Coast, Guinea-Bissau, Mali, Niger, Senegal, and Togo.
CFA stands for Communauté Financière d'Afrique (Financial Community
of Africa). The currency is issued by the BCEAO (Banque Centrale des
États de l'Afrique de l'Ouest, the Central Bank of the West African States),
located in Dakar. —Trans.

216 *FOB (free on board)* FOB is a system that specifies whether the buyer or the
seller pays for shipment and loading costs, and/or where responsibility for
goods is transferred from buyer to seller, and therefore also which party is
liable for goods lost or damaged in transit. —Trans.

216 *spot market* A spot market, also called a cash market, is a public financial
market on which commodities or financial markets are traded for immediate delivery. By contrast, a futures market handles trades whose delivery is due at a later date. The spot market today functions largely over the
Internet. —Trans.

216 *This is how . . . economy* See "L'inquiétante volatilité des prix des matières
premières agricoles," *Le Monde*, January 11, 2011, www.lemonde.fr
/planete/article/2011/01/11/l-inquietante-volatilite-des-prix-des-matieres
-premieres-agricoles_1463798_3244.html. See also the World Bank,
Poverty Reduction and Equity Group, "Food Price Watch," February
2011, siteresources.worldbank.org/INTPOVERTY/Resources/335642
-1210859591030/Food_Price_Watch_Feb2011.pdf.

216 *The World Bank's . . . (12%)* World Bank, "Food Price Watch," 1 (see the
preceding note).

216 *Higher global wheat... (16%)*   Ibid., 2.

217 *Maize prices have... prices*   Ibid., 2.

217 *Domestic rice prices... poor*   Ibid., 4. See also Jean-Christophe Kroll and Aurélie Trouvé, "G20 et sécurité alimentaire: la vanité des discours," *Le Monde*, February 28, 2011, www.lemonde.fr/idees/article/2011/02/28/g20-et-securite-alimentaire-la-vanite-des-discours_1486039_3232.html.

218 *There would not... products*   Isabelle Hachey, "La spéculation au coeur de la crise alimentaire," interview with Olivier De Schutter, LaPresse.ca, October 16, 2010.

218 *The impact of... control*   UN Conference on Trade and Development (UNCTAD), *Trade and Development Report 2008: Commodity Prices, Capital Flows and the Financing of Investment* (Geneva, 2008), II.

219 *Our members represent... US$5 billion*   World Economic Forum, Frequently Asked Questions, www.weforum.org/faq.

219 *"What kind of civilization... eat?"*   See Chalmin, *Le monde a faim*, 52 (see note to p. 211).

220 *"We have to snatch . . . speculators"*   Heiner Flassbeck, "Rohstoffe den Spekulanten entreissen," *Handelsblatt*, Düsseldorf, February 11, 2011. [My translation from the original German here is via Jean Ziegler's French, since the *Handelsblatt* has no online presence. —Trans.]

220 *The proposal by Fassbeck . . . 2009*   See Robles, Torero, and Braun, *When Speculation Matters* (see note to p. 208).

220 *But what is lacking... it*   In the United States there is an agency appointed to regulate speculation in food commodities, the U.S. Commodity Futures Trading Commission; the commission has proven to be exceptionally ineffectual.

27. GENEVA, WORLD CAPITAL OF AGRI-FOOD SPECULATORS

221 *"is equally inseparable... lobby"*   Marc Roche, "Haro sur les spéculateurs fous!" *Le Monde*, January 30, 2011, available to subscribers or for purchase at www.lemonde.fr/cgi-bin/ACHATS/acheter.cgi?offre=ARCHIVES&type_item=ART_ARCH_30J&objet_id=1147156&xtmc=speculateurs&xtcr=1.

222 *In this sector... London*   In 2009, Prime Minister Gordon Brown took severe measures to limit the bonuses, stock options, raises, and other premiums added to the exorbitant incomes of hedge fund managers: anyone paid more than 200,000 pounds (about $325,000) annually is taxed at the rate of 50 percent on the excess.

222 *Jabre Capital Partners*   See Marc Roche's profile of Jabre, "Le retour en force de Philippe Jabre, roi des hedge funds," *Le Monde*, April 26, 2011, available to subscribers or for purchase at www.lemonde.fr/cgi-bin/ACHATS/acheter.cgi?offre=ARCHIVES&type_item=ART_ARCH_30J&objet_id=1155273&xtmc=philippe_jabre&xtcr=1.

222 *The volume of business... 2010*   See "Genève, paradis du négoce," *La Liberté*, Fribourg, March 25, 2011, available to subscribers or for purchase at www.laliberte.ch/economie/geneve-paradis-du-negoce.

222 *In addition... law*   See the investigative report by Elisabeth Eckert, "1500 milliards de francs, au moins, échappent à tout contrôle en Suisse," *Le*

Matin Dimanche, April 2, 2011,,,m,,,, new.lematin.ch/actu/economie/1500 -milliards-de-francs-au-moins-%C3%A9chappent-%C3%A0-tout-con tr%C3%B4le-en-suisse.

224 *"We don't regulate . . . so"* See Eckert, "1500 milliards de francs" (see the preceding note). [See also Pierre-François Besson, "Financial Regulator Says Crisis Not Over," interview with Anne Héritier Lachat, trans. Clare O'Dea, Swissinfo.ch, March 23, 2011, www.swissinfo.ch/eng/business /Financial_regulator_says_crisis_not_over.html?cid=29810118). —Trans.]

224 *"How can it be . . . oversight?"* See Eckert, "1500 milliards de francs" (see note to p. 222).

224 *(JetFin is a . . . industry)* For JetFin's own version of its activities, see www .jetfin.com/english. —Trans.

224 *"Agriculture today is . . . markets"* The quote here is my translation from a JetFin brochure text quoted by Jean Ziegler, which is not available online. See also "Why Attend?" at www.jetfin.com/agro2011-geneva/eventinfo _en.php: "This event brings together investors and fund managers to examine investment strategies and risks in the areas of soft commodities, farming, carbon, timber, water. JetFin AGRO 2011 conference is a unique event, gathering top fund managers deploying winning investment strategies in agriculture. Leading fund managers present and explain their investment strategies. International investors searching hard assets and real alpha in natural resources and agriculture can meet face to face with some of the best fund managers in the field. Fund of funds managers and investors share their views on current strategic and tactical allocations in soft commodities and agriculture." —Trans.

224 *Two powerful NGOs . . . government* For more information about Fastenopfer, see www.fastenopfer.org/sites/home/index.html?lang2=en; for Bread for All, see www.ppp.ch/en/english. The June 28, 2010, letter appears not to be available online. However, a letter addressed to the Geneva government on May 25, 2011, protesting the following JetFin conference in 2011, presumably in similar terms, and signed by twenty-two NGOs led by Fastenopfer and Bread for All, is available at www.google.com/url?sa =t&rct=j&q=&esrc=s&source=web&cd=1&ved=0CGMQFjAA&url= http%3A%2F%2Fwww.agrisodu.ch%2Findex.php%3Foption%3Dcom _docman%26task%3Ddoc_download%26gid%3D299%26Itemid%3D& ei=QDCsT6C-EIPC0QH8s_j6Dw&usg=AFQjCNEQQ-QgO1FbEkQ6 Ycs5hGJoK9ZntA&sig2=emSf40cZY5Xzm1dMKjnduw. —Trans.

28. LAND GRABS AND THE RESISTANCE OF THE DAMNED

226 *But even before . . . (988,000 acres)* Marc Guéniat, "Le jeune Sud-Soudan brade déjà ses terres agricoles—Une société texane a obtenu le contrat du siècle: 600,000 hectares pour 25,000 dollars, avec agriculture et pétrole à la clé," *La Tribune de Genève,* June 9, 2011, archives.tdg.ch /jeune-sud-soudan-brade-terres-agricoles-2011-06-09.

226 *In Switzerland . . . clientele* Yvan Maillard Ardenti, "Accaparement des terres et flux financiers internationaux: L'Implication du secteur financier Suisse," in *L'accaparement des terres: La course aux terres aggrave la faim dans le*

monde, ed. Pascale Schnyder and Ester Wolf (Lausanne: Pain pour le Prochain and Action de Carême [Bread for All and Fastenopfer], in the series Collection Repères, January 2010), 16–17, www.ppp.ch/fileadmin /francais/Politique_developpement/Reperes%20et%20publications /Reperes-1-2010_01.pdf; also available in German. [This report is not available in English, but see Anh-Nga Tran Nguyen, "Global Land Grabbing: Issues and Solutions," Bread for All, September 2010, www .brotfueralle.ch/fileadmin/deutsch/2_Entwicklungpolitik_allgemein/A _Recht_auf_Nahrung/Global%20land%20grabbing%20by%20ATN .pdf. —Trans.]

227 *The Sarasin and Pictet . . . Russia* Ibid., 17.

227 *In the context . . . consequences* Miges Baumann, "La soif de terres aggrave la faim dans le monde" in Schnyder and Wolf, *L'accaparement des terres*, 7 (see note to p. 226).

228 *We might consider . . . effort* Alexandre Vilgrain, "Jouons Collectifs!" editorial, *La Lettre du CIAN*, November–December 2010, 1, www.cian.asso .fr/cianweb/cianweb-img.nsf/FindIMG/3.-La%20Lettre%20du%20 CIAN%20Nov-Dec%202010.pdf/$FILE/La%20Lettre%20du%20 CIAN%20Nov-Dec%202010.pdf.

229 *World Social Forum* For more on the World Social Forum 2011, see fsm2011 .org. —Trans.

229 *Cameroon, which is . . . continent* See Transparency International's annually updated Corruption Perceptions Index; the 2011 index is available at cpi .transparency.org/cpi2011/results.

229 *This time, SOSUCAM . . . annually* See "Cameroun: Somdiaa sucre les droits," Appels Urgents 341, Peuples Solidaires/ActionAid, February 12, 2010, www.peuples-solidaires.org/341-cameroun-somdiaa-sucre-les-droits. [Basic information about Peuples Solidaires is available in English at www .peuples-solidaires.org/welcome. —Trans.]

230 *According to the resistors . . . processing* Ibid.

230 *"human values are . . . Group"* As recently as mid-2012, SOMDIAA included this statement on an "About Us" page at its website: www.somdiaa.com /en/groupe/nous-connaitre. The slogan was widely quoted and excoriated in the online pro–farmers' rights press. SOMDIAA now has a much more elaborate statement of its values at www.somdiaa.com/groupe/les-valeurs (in French only), which opens with the claim that as a "family-owned company, SOMDIAA has always supported business practices based on respect for fundamental human values, which are shared with all of our partners," and continues at length in a similar vein. —Trans.

231 *The majority of . . . less* Ester Wolf, "Spéculation foncière au Bénin au détriment des plus pauvres," in Schnyder and Wolf, *L'accaparement des terres*, 20–22 (see note to p. 226).

231 *"the current prices . . . forever"* Ibid., 21.

232 *Euro RSCG* In the fall of 2012, Euro RSCG Worldwide was renamed Havas Worldwide. —Trans.

232 *However, Yayi was . . . voters)* Philippe Perdrix, "Bénin: Boni Yayi, vainqueur par K.O., investi pour un second quinquennat," *Jeune Afrique*, April 8, 2011, www.jeuneafrique.com/Article/ARTJAJA2620p024-026.xml0. [The op-

position candidates in Benin alleged massive electoral fraud, but the election results were upheld by the country's constitutional court. See Dwyer Ace, "Benin Court Confirms Presidential Election Results Amid Fraud Allegations," Jurist Paper Chase Newsburst, March 21, 2011, jurist.org /paperchase/2011/03/benin-court-confirms-presidential-election-results -amid-fraud-allegations.php. —Trans.]

232 *"While local small farmers . . . fallow"* Quoted in Wolf, "Spéculation foncière," 20 (see notes to pp. 226 and 231).

232 *ROPPA* For more on ROPPA, see p. 49.

232 *In 2008 . . . government* See Philippe Bernard, "La crise en Libye remet en cause la cession d'une vaste zone rizicole au Mali," *Le Monde*, April 1, 2011, available to subscribers or for purchase at www.lemonde.fr/cgi-bin /ACHATS/acheter.cgi?offre=ARCHIVES&type_item=ART_ARCH _30J&objet_id=1153008&xtmc=malibya&xtcr=1.

233 *"The Libyans behave . . . here"* Ibid.

233 *"run on agricultural . . . imprisoned"* Ibid.

233 *"the necessity of . . . area"* Ibid.

29. THE COMPLICITY OF THE WESTERN STATES

239 *Considering that recent . . . land* World Social Forum 2011, Dakar Appeal Against the Land Grab; the text of the appeal is available at many websites, including www.petitiononline.com/dakar/petition.html.

EPILOGUE

242 *But where there . . . grows* Hölderlin's lines, which come from one of his most famous poems, read in German: "Wo aber Gefahr ist, wächst / Das Rettende auch." The translation in the epigraph is my own. —Trans.

244 *"Earth provides enough . . . greed"* This proverb has entered the realm of Gandhi folklore and is attributed to the Mahatma in many different and slightly erroneous versions. The original source is a firsthand witness, Pyarelal, in a chapter titled "Towards New Horizons" in part two of his *Mahatma Gandhi: The Last Phase* (1958 and later editions); for an authoritative account, with interesting observations on Gandhi's evolving and prescient views on food, hunger, and ecology, see Y.P. Anand and Mark Lindley, "Gandhi on Providence and Greed," available at independent.academia .edu/MarkLindley/Papers/255823/Gandhi_on_providence_and_greed. The full context of Pyarelal's account of Gandhi's remark speaks even more potently to Jean Ziegler's work:

> In addition to the economic and the biological, there is another aspect of man's being that enters into [human] relationships with nature, namely the spiritual. When the balance between the spiritual and the material is disturbed, sickness results.
>
> "Earth [*pritvi*, the world] provides enough to satisfy every man's need but not for every man's greed," said Gandhiji. So long as we cooperate with the cycle of life, the soil renews its fertility indefinitely and provides health, recreation, sustenance and peace to those who

depend on it. But when the "predatory" attitude prevails, nature's balance is upset and there is an all-round biological deterioration. —Trans.

244　*Eric Toussaint . . . Germany*　Eric Toussaint, Damien Millet, and Daniel Munevar, "Les chiffres de la dette 2011," CADTM (Comité pour l'Abolition de la Dette du Tiers-monde), Liège, April 19, 2011, www.cadtm.org/IMG /pdf/Les_chiffres_de_la_dette_2011_DEf.pdf; also available in Spanish and Portuguese (but not in English) at www.cadtm.org/Les-chiffres-de-la -dette-2011. See also, by the same authors, with others, *La Dette ou la Vie* (Brussels and Liège: ADEN and CADTM, 2011); and Capgemini and Merrill Lynch Global Wealth Management, *World Wealth Report 2011*, www .us.capgemini.com/services-and-solutions/by-industry/financial-services /solutions/wealth/worldwealthreport.

245　*"In the field . . . important"*　Amartya Sen, "Food, Economics and Entitlements," WIDER Working Paper 1, United Nations University World Institute for Development Economics Research, February 1986, 4, www .wider.unu.edu/publications/working-papers/previous/en_GB/wp-01 /_files/82530785773748791/default/WP1.pdf.

246　*First, by combating . . . bring*　See the classic treatment of this subject by Georg Cremer, *Corruption and Development Aid: Confronting the Challenges* (London: Lynne Rienner, 2008).

247　*"Pessimism of the . . . will"*　Antonio Gramsci, in a letter from prison to his brother Carlo, dated December 19, 1929, in Gramsci, *Selections from the Prison Notebooks*, ed. Quintin Hoare and Geoffrey Nowell Smith (London: Electric Book Company, 2001; New York: International Publishers, 1971), 395. While Gramsci made this maxim, and many variants of it, famous, Hoare and Smith, among others, attribute it originally to Romain Rolland. —Trans.

248　*Almost half of . . . life*　La Via Campesina, "Declaration of Rights of Peasants—Women and Men," March 2009, viacampesina.net/down loads/PDF/EN-3.pdf. [A better translation of *campesinos/campesinas* (or the French *paysans/paysannes*) than "peasants—women and men" would be "small farmers"—which is the term I have adopted throughout this book. —Trans.]

248　Sólo le pido . . . *enough*　*"Sólo le pido a Dios,"* words and music by Léon Gieco (1978); translation mine. —Trans.